GHRELIN

ENDOCRINE UPDATES
Shlomo Melmed, M.D., Series Editor

J.A. Fagin (ed.): Thyroid Cancer. 1998.ISBN: 0-7923-8326-5
J.S. Adams and B.P. Lukert (eds.): Osteoporosis: Genetics,
Prevention and Treatment. 1998. ISBN: 0-7923-8366-4.
B.-Å. Bengtsson (ed.): Growth Hormone. 1999. ISBN: 0-7923-8478-4
C. Wang (ed.): Male Reproductive Function. 1999. ISBN: 0-7923-8520-9
B. Rapoport and S.M. McLachlan (eds.): Graves' Disease:
Pathogenesis and Treatment. 2000. ISBN: 0-7923-7790-7.
W. W. de Herder (ed.): Functional and Morphological Imaging
of the Endocrine System. 2000. ISBN 0-7923-7923-9
H.G. Burger (ed.): Sex Hormone Replacement Therapy. 2001.
ISBN 0-7923-7965-9
A. Giustina (ed.): Growth Hormone and the Heart. 2001.
ISBN 0-7923-7212-3
W.L. Lowe, Jr. (ed.): Genetics of Diabetes Mellitus. 2001.
ISBN 0-7923-7252-2
J.F. Habener and M.A. Hussain (eds.): Molecular Basis of Pancreas
Development and Function. 2001. ISBN 0-7923-7271-9
N. Horseman (ed.): Prolactin. 2001. ISBN 0-7923-7290-5
M. Castro (ed.): Transgenic Models in Endocrinology. 2001
ISBN 0-7923-7344-8
R. Bahn (ed.): Thyroid Eye Disease. 2001. ISBN 0-7923-7380-4
M.D. Bronstein (ed.): Pituitary Tumors in Pregnancy
ISBN 0-7923-7442-8
K. Sandberg and S.E. Mulroney (eds.): RNA Binding Proteins:
New Concepts in Gene Regulation. 2001. ISBN 0-7923-7612-9
V. Goffin and P. A. Kelly (eds.): Hormone Signaling. 2002
ISBN 0-7923-7660-9
M. C. Sheppard and P. M. Stewart (eds.): Pituitary Disease. 2002
ISBN 1-4020-7122-1
N. Chattopadhyay and E.M. Brown (eds.): Calcium-Sensing Receptor.
2002. ISBN 1-4020-7314-3
H. Vaudry and A. Arimura (eds.): Pituitary Adenylate Cyclase-
Activating Polypeptide. 2002. ISBN: 1-4020-7306-2
Gaillard, Rolf C. (ed.): The ACTH AXIS: Pathogenesis, Diagnosis
and Treatment. 2003. ISBN 1-4020-7563-4
Beck-Peccoz, Paolo (ed.): Syndromes of Hormone Resistance on the
Hypothalamic-Pituitary-Thyroid Axis. 2004. ISBN 1-4020-7807-2
Ghigo, Ezio (ed.): Ghrelin. 2004. ISBN 1-4020-7770-X
Srikant, Coimbatore B. (ed.): Somatostatin. 2004.
ISBN 1-4020-7799-8

GHRELIN

edited by

Ezio Ghigo

with the collaboration of

Andrea Benso *and* Fabio Broglio

Division of Endocrinology and Metabolism
Department of Internal Medicine
University of Turin, Italy

SPRINGER SCIENCE+BUSINESS MEDIA, LLC

Library of Congress Cataloging-in-Publication Data

A C.I.P. Catalogue record for this book is available
from the Library of Congress.

Ghrelin
Ezio Ghigo
ISBN 978-1-4757-8841-9   ISBN 978-1-4020-7971-9 (eBook)
DOI 10.1007/978-1-4020-7971-9

Copyright © 2004 by Springer Science+Business Media New York
Originally published by Kluwer Academic Publishers in 2004
Softcover reprint of the hardcover 1st edition 2004

All rights reserved. No part of this work may be reproduced, stored in a retrieval
system, or transmitted in any form or by any means, electronic, mechanical,
photocopying, microfilming, recording, or otherwise, without the written permission
from the Publisher, with the exception of any material supplied specifically for the
purpose of being entered and executed on a computer system, for exclusive use by
the purchaser of the work.

Permission for books published in Europe: permissions@wkap.nl
Permissions for books published in the United States of America: permissions@wkap.com

Printed on acid-free paper.

*The Publisher offers discounts on this book for course use and bulk purchases.
For further information, send email to <Laura.Walsh@wkap.com>.*

Contents

List of Contributors

Alex R.T. Bailey — Huffington Center on Aging and Departments of Molecular and Cellular Biology and Medicine, Baylor College of Medicine, Houston, Texas, USA

Marie Hélène Bassant — Institut National de la Santé et de la Recherche Médicale INSERM U.549, IFR Broca Ste-Anne, Paris, France

Simonetta Bellone — Unit of Pediatrics, Department of Medical Sciences, University of Piemonte Orientale, Novara, Italy

Marie Thèrése Bluet-Pajot — Institut National de la Santé et de la Recherche Médicale INSERM U.549, IFR Broca Ste-Anne, Paris, France

Gianni Bona — Unit of Pediatrics, Department of Medical Sciences, University of Piemonte Orientale, Novara, Italy

Cyril Y. Bowers — Department of Medicine, Endocrinology and Metabolism Section, Tulane University Health Sciences Center, New Orleans, Louisiana, USA

Fabio Broglio — Division of Endocrinology and Metabolism, Department of Internal Medicine, University of Turin, Turin, Italy

Felipe F. Casanueva — Department of Medicine, University of Santiago de Compostela, Santiago de Compostela, Spain

Tamara Castañeda — Department of Pharmacology, German Institute of Human Nutrition, Potsdam-Rehbrücke, Germany

Fernando Cordido — Department of Medicine and Physiology, University of Coruna, La Coruna, Spain

David E. Cummings — Department of Medicine, Division of Metabolism, Endocrinology and Nutrition, University of Washington, VA Puget Sound Health Care System, Seattle, WA, USA

Roland Dardennes — Institut National de la Santé et de la Recherche Médicale INSERM U.549 and CMME, IFR Broca Ste-Anne, Paris, France

Yukari Date — Department of Internal Medicine, Miyazaki Medical College, Miyazaki and National Cardiovascular Center Research Institute, Suita, Osaka, Japan

Carlos Diéguez — Department of Physiology, University of Santiago de Compostela, Spain

Jacques Epelbaum — Institut National de la Santé et de la Recherche Médicale INSERM U.549, IFR Broca Ste-Anne, Paris, France

Bruno Estour — Hôpital Bellevue, St Etienne, France

Christine Foulon — CMME, IFR Broca Ste-Anne, Paris, France

Alessandra Gambineri — Endocrinology Unit, Department of Internal Medicine and Gastroenterology, Center of Applied Biomedical Research, S.Orsola-Malpighi Hospital, Bologna, Italy

Carlotta Gauna — Department of Internal Medicine, Section of Endocrinology, Erasmus MC, Rotterdam, The Netherlands

Corrado Ghè — Department of Anatomy, Pharmacology, and Forensic Medicine, University of Turin, Turin, Italy

Ashley B. Grossman — Department of Endocrinology, St. Bartholomew's Hospital, London, UK

Ezio Ghigo — Division of Endocrinology and Metabolism, Department of Internal Medicine, University of Turin, Turin, Italy

Julien Daniel Guelfi — CMME, IFR Broca Ste-Anne, Paris, France

Hiroshi Hosoda — Department of Biochemistry, National Cardiovascular Center Research Institute, Suita, Osaka and Translational Research Center, Kyoto University Hospital, Kyoto, Japan

Jörgen Isgaard — Research Center for Endocrinology and Metabolism, Sahlgrenska University Hospital, Göteborg, Sweden

Inger Johansson — Research Center for Endocrinology and Metabolism, Sahlgrenska University Hospital, Göteborg, Sweden

Hiroyuki Kaiya — Department of Biochemistry, National Cardiovascular Center Research Institute, Osaka, Japan

Kenji Kangawa — Department of Biochemistry, National Cardiovascular Center Research Institute, Osaka and Translational Research Center, Kyoto University Hospital, Kyoto, Japan

Masayasu Kojima — Molecular Genetics, Institute of Life Science, Kurume University, Kurume, Fukuoka, Japan

Márta Korbonits — Department of Endocrinology, St. Bartholomew's Hospital, London, UK

Yves Le Bouc — INSERM U.515, Hôpital St-Antoine, Paris, France

Hisayuki Matsuo — Department of Internal Medicine, Miyazaki Medical College, Miyazaki, Japan,

Giampiero Muccioli — Department of Anatomy, Pharmacology, and Forensic Medicine, University of Turin, Turin, Italy

Masamitsu Nakazato — Department of Internal Medicine, Miyazaki Medical College, Miyazaki, Japan

Joost Overduin — Department of Medicine, Division of Metabolism, Endocrinology And Nutrition, University of Washington, VA Puget Sound Health Care System, Seattle, WA, USA

Uberto Pagotto — Endocrinology Unit, Department of Internal Medicine and Gastroenterology, Center of Applied Biomedical Research, S.Orsola-Malpighi Hospital, Bologna, Italy

Mauro Papotti — Department of Biomedical Sciences & Oncology, University of Turin and San Luigi Hospital, Turin, Italy

Antonia Paschali	Huffington Center on Aging and Departments of Molecular and Cellular Biology and Medicine, Baylor College of Medicine, Houston, Texas, USA
Renato Pasquali	Endocrinology Unit, Department of Internal Medicine and Gastroenterology, Center of Applied Biomedical Research, S.Orsola-Malpighi Hospital, Bologna, Italy
Frédérique Poindessous-Jazat	Institut National de la Santé et de la Recherche Médicale INSERM U.549, IFR Broca Ste-Anne, Paris, France
Vera Popovic	Institute of Endocrinology, University of Belgrade, Belgrade, Serbia
Anna Rapa	Unit of Pediatrics, Department of Medical Sciences, University of Piemonte Orientale, Novara; Italy
Marie Christine Rio	IGBMC, CNRS/INSERM U.184/ULP, Illkirch, France
Roy G. Smith	Huffington Center on Aging and Departments of Molecular and Cellular Biology and Medicine, Baylor College of Medicine, Houston, Texas, USA
Yuxiang Sun	Huffington Center on Aging and Departments of Molecular and Cellular Biology and Medicine, Baylor College of Medicine, Houston, Texas, USA
Gloria S. Tannenbaum	Departments of Pediatrics and Neurology and Neurosurgery, McGill University, Montreal, Quebec, Canada
Elena Tarabra	Department of Anatomy, Pharmacology, and Forensic Medicine, University of Turin, Turin, Italy
Åsa Tivesten	Research Center for Endocrinology and Metabolism, Sahlgrenska University Hospital, Göteborg, Sweden
Virginie Tolle	Institut National De La Santé et de la Recherche Médicale INSERM, Paris, France
Catherine Tomasetto	IGBMC, CNRS/INSERM U.184/ULP, Illkirch, France
Matthias H. Tschöp	Department of Pharmacology, German Institute of Human Nutrition, Potsdam-Rehbrücke, Germany
Aart Jan van der Lely	Department of Internal Medicine, Section of Endocrinology, Erasmus MC, Rotterdam, The Netherlands
Valentina Vicennati	Endocrinology Unit, Department of Internal Medicine and Gastroenterology, Center of Applied Biomedical Research, S.Orsola-Malpighi Hospital, Bologna, Italy
Marco Volante	Department of Biomedical Sciences & Oncology, University of Turin, Turin, Italy
Philippe Zizzari	Institut National de la Santé et de la Recherche Médicale INSERM U.549, IFR Broca Ste-Anne, Paris, France

Preface

It is my pleasure and honor to welcome the reader of this book dedicated to ghrelin, a new gastric hormone.

Ghrelin is the result of a story of reverse pharmacology which started more than 20 years ago with the discovery of synthetic, non natural Growth Hormone Releasing Peptides (GHRP). Then, the following milestones of the story were the discovery of non peptidyl GH Secretagogues (GHS) and that of the GHS receptor.

The ghrelin story was at first closely related to the regulation of GH secretion but we now know ghrelin is much more than simply a natural GHS. At present, the orexigenic action of ghrelin is attracting major interest, having opened new understandings in the regulation of appetite and food intake. But again, ghrelin is also much more than just a natural orexigenic factor. We still don't know enough about the physiology and pathophysiology of ghrelin but it is already clear that this hormone plays other remarkable central and peripheral actions including influence on behaviour, hypothalamus-pituitary functions, activities and secretion of peripheral endocrine glands, metabolic balance, gastroenteropancreatic functions, cardiovascular functions, normal and neoplastic cell proliferation. Thus, the ghrelin story clearly appears no longer just as a milestone in neuroendocrinology but as an important aspect of internal medicine from basic to clinical aspects.

I am indebted to all the Authors who contributed chapters of this book; herein you can find almost all the scientists who provided major contributions to this new field. In particular, I would like to dedicate this book to those who more than any other made this story real: Cyril Bowers who, with his group, discovered GHRP, Roy Smith who, with his group, generated non peptidyl GHS and discovered the GHS receptor, Kenij Kangawa who, with his group, discovered ghrelin.

It is my hope that this book will provide readers with an exhaustive picture of present knowledge in the ghrelin field. Given the impressive speed in the growing body of information and discoveries, it is clear that this book will soon be obsolete but nevertheless it will remain a good basis for better understanding the novelties awaiting us in the near future.

Ezio Ghigo

Chapter 1

HISTORICAL MILESTONES

Cyril Y. Bowers

Endocrine Section, Tulane University School of Medicine, New Orleans, Louisiana 70112-2699, USA

Abstract: Three premier core historical milestones, the first unnatural Growth Hormone Releasing Peptide (GHRP) in 1976, the GHRP/Growth Hormone Secretagogue (GHS) receptor in 1996 and the natural GHRP hormone ghrelin in 1999 form the foundation of a new physiological system in animals and humans. The talents of many investigators worldwide on the actions, scope and various biological actions of the ghrelin system have been defined and redefined. This has been accomplished with many different unique and novel unnatural GHRPs initially and increasingly with the new ghrelin hormone. Results obtained with the unnatural GHRPs and ghrelin have been essentially the same. The availability of natural ghrelin has significantly modified and enlarged the scope of the endogenous system. The increased secretion of growth hormone via a direct combined action on the hypothalamus and pituitary as well as food intake via a hypothalamic action have been firmly established. From the beginning of the GHRP/receptor/ghrelin story, surprising results have been and continue to be a hallmark. The primary origin of ghrelin from the stomach indeed was most unexpected and conceptually reorienting. Already evidence supports physiological and pathophysiological roles involving the ghrelin system and new GHRP/ghrelin strategies, approaches and techniques. Differentiating the physiological from the pharmacological actions of GHRP/ghrelin will be an immediate future challenge. So many unnatural and natural disparate findings have come together in such a complementary and novel way that the unexpected appears likely to persist in the future.

Key words: GHRP, GHRP/GHS/ghrelin receptor

1.   INTRODUCTION

The unnatural artificial origin of Growth Hormone Releasing Peptide (GHRP) and the successful but also surprising transformation all the way into a new natural biological system in both animals and humans underscore a most special aspect of the story (1). The talents of many investigators worldwide convincingly, meaningfully and excitedly have made this story possible.

To lessen the controversy of the milestone selection process, an emphasis has been given to evolutionary foreverness and obvious fundamental importance rather than originality and novelty although each exists and is involved in the still developing and establishing futuristic findings such as receptor subtypes, unique cardiovascular actions and paracrine-autocrine actions. Over the period between 1976 and 1999 the three major milestones of the GHRP story emerged. The first core premier historical milestone was the initial unnatural GHRP that evolved in 1976, its receptor was revealed in 1996 as the second milestone and the elegant climax of this triumvirate core finally appeared as the natural hormone, ghrelin, in 1999. Superimposed on this triumvirate core of milestones has been a number of unique and novel biological actions of GHRPs and, more recently, ghrelin.

Most fortunately the diverse basic and clinical results of the unnatural GHRP-Growth Hormone Secretagogue (GHS) amassed over the past 25 years so far have paralleled those obtained with the natural GHRP designated ghrelin in animals as well as in humans (2). Importantly, most of the time the results, past and present, of the unnatural GHRPs and natural ghrelin can be directly interrelated. Nevertheless, surprises have consistently been part of the GHRP saga.

Especially unexpected has been the wide range of chemistries of these GHSs and their high affinity and efficacy for the same singular natural receptor. Natural GHRP, ghrelin, was first conceived as anatomically originating from the hypothalamus and functioning as an additional hypothalamic hypophysioptropic hormone involved in the physiological regulation of growth hormone (GH) secretion. Instead, natural ghrelin originates from the stomach and has become a gastric hormone with some evidence of anatomically originating in the pituitary but very little if any from the hypothalamus. The implication of these anatomical ghrelin findings has and continues to incite considerable speculation about the physiological role and scope of ghrelin and how this all occurs.

To impart a brief and arbitrary commentary on the historical milestones of the GHRPs, GHS receptor (GHS-R) and ghrelin story, salient points will be presented from three viewpoints of the chemistry, receptor and biological actions. Indeed these triumviral milestones have come together as a solid foundation for readily expecting new peptides or peptidomimetics that meaningfully will relate to health and disease. Especially challenging in the future will be the decision of what and when findings of this dynamic and expanding story sufficiently command the status of a milestone.

2. CHEMISTRY

Premier chemical historical milestones occurred in 1976 with the first unnatural GHRP and in 1999 with the natural hormone designated ghrelin (3,4). The first historical milestone was chemically unrefined but nevertheless its importance was underscored by specific GH releasing activity via a direct pituitary action and this was when the GHRP saga all began. The 1999 premier historical milestone ghrelin was chemically refined and was a long-anticipated finale by many investigators.

The first synthetic unnatural GHRP that we developed was designated $DTrp^2$. This pentapeptide, $TyrDTrpGlyPheMetNH_2$ was a derivative of the natural opiate Met enkephalin pentapeptide (TyrGlyGlyPheMet), isolated in 1975 by Hughes and coworkers (5) which also released GH but, as eventually determined, by a hypothalamic rather than a pituitary anatomic site of action. $DTrp^2$ was of low potency and was active only *in vitro* but not *in vivo*. Nevertheless, this was the first peptide of known chemical sequence that acted directly on the pituitary to release GH. $DTrp^2$ was specific in action and had no opiate activity.

Of chemical significance are the GHRPs which are classified into 4 groups according to the number, position and chirality of the Trp residue (1980). They are $TyrDTrp^2GlyPheMetNH_2$ ($DTrp^2$), $TyrAlaDTrp^3PheMet$ NH_2 ($DTrp^3$), $TyrDTrp^2DTrp^3PheMetNH_2$ ($DTrp^{2,3}$) and $TyrDTrp^2AlaTrp$ $DPheNH_2$ ($DTrp^2,LTrp^4$). Notable is that each of these four chemical general classes of GHRP have been utilized as templates for new GHRPs with high potency by ourselves and other investigators. GHRP-6, $HisDTrpAlaTrpDPheLysNH_2$, became the second chemical milestone because it was active both *in vitro* and *in vivo* and was hypothesized in 1984 to reflect the GH releasing activity of a new hormone different form Growth Hormone Releasing Hormone (GHRH). In 1989, GHRP-6 became the first GHRP to release GH in humans. Additionally, our low potency GHRP receptor antagonists $DLys^3$-GHRP-6 ($HisDTrpDLys^3$ $TrpDPheLysNH_2$, 1980) and the $[DArg^1DPhe^5DTrp^{7,9}Leu]$-Substance P antagonist (1993) represent two of the very few reported that inhibit the GH response to GHRP and ghrelin in rats.

Particularly remarkable is the broad range of chemistries of the GHRPs developed by eight different pharmaceutical groups (1990-1999) as well as our group. They consist of low molecular weight peptides, partial peptides and nonpeptides. In spite of very different chemistries, many GHRPs are of high potency and nearly all have the same biological profile. Theoretical projections of this unusual chemical diversity, interestingly, could mean potent GHRPs of diverse chemistry may dock at different sites on the receptor and still activate the receptor. Sometimes this may be activation of a different, additional and/or modified intracellular signal transduction pathway.

The third and final core premier historical milestone, the outstanding accomplishment of Kojima, Hosoda, Date, Nakazato, Matsuo and Kangawa in 1999, was the isolation and identification of ghrelin that subsequently made possible the linking of unnatural GHRP and natural ghrelin (4,6). Their chemical differences are amazing especially in the context that both the unnatural and the natural molecules bind with high affinity to the same receptor and produce the same biological actions. Furthermore, the attachment and obligate biological activity requirement of the octanoyl adduct covalently linked to the hydroxyl group of the side chain of serine 3 via an ester bond adds to the amazement.

Octanoyl acylation of the ghrelin molecule is a first in bioactive peptides and it is astounding that the 28 amino acid non-octanoylated peptide is biologically inactive in spite of the large number and diversity of chemical functional groups of the peptide and the chemical inertness of the hydrophobic octanoic addition. Fortuitously, desoctanoylated ghrelin is biologically inactive because it is 3-4 or more times higher in concentration than the bioactive octanolated ghrelin in peripheral plasma. If desoctanoylated ghrelin were bioactive, either as an antagonist or agonist, this would significantly complicate the interpretation of the octanoylated ghrelin action and role. The fact that octanoylation determines whether ghrelin is bioactive suggests octanoylation mainly determines the active conformation of the molecule rather than just modulating the ghrelin pharmacokinetics and pharmacodynamics. Already highly homologous human ghrelin molecules have been cloned by Kangawa and coworkers from rat, mouse, dog, pig, sheep, cow, chicken, bullfrog, and eel. The chemical acyl acid hallmark as an octanoic or decanoic addition has been consistently part of the ghrelin molecule from all species.

3.   RECEPTOR

Starting with the new GHRP/GHS-R in 1996 and ending with the new GHRP/GHS hormone ghrelin in 1999, the unnaturalness of the GHRPs at last became scientifically and definitively natural. Identification of the natural receptor and hormone has become the foundation of an entirely new and unique biological hormonal system. Unnatural GHRPs now can be interpreted more meaningfully in terms of physiology and pathophysiology. The widespread expression of this G-protein coupled seven transmembrane receptor by a highly conserved gene in the human, chimpanzee, pig, cow, rat and mouse adds to its biological importance. Identification of the homolog of the human GHRP/GHS-R in the puffer fish genome imparts a 400 million year old evolutionary foreverness.

It was the Merck group in 1992 under the direction of Roy Smith, one of the early believers in GHRPs, that made the GHRP saga even more exciting, with the development of the potent peptidomimetic GHRP/GHS, MK-0677. This peptidomimetic was active orally with high bioavailability and a longer duration of action. In 1996 Howard and coworkers, also at Merck, made another outstanding contribution by cloning the gene for the GHRP/GHS-R, a new G-protein coupled seven transmembrane receptor (7,8). A receptor specifically different from GHRH was strongly supported by the differences in the GH releasing activity of GHRP versus GHRH as well as the pituitary intracellular signal pathways of these two peptides. GHRH acts via the adenyl cyclase protein kinase A pathway and GHRP/ghrelin via the phospholipase protein C pathway. This 1996 receptor identification is considered the second of 3 core premier historical milestones.

The selected site-directed mutation studies in 1998 by Feighner and coworkers (9) at Merck revealed sites on the transmembrane domains 2,3 and 5,6 of the ghrelin/GHS-R that affected binding and activation. By mutating glutamic acid to glutamine at position 124 in transmembrane domain 3, a nonfunctional receptor resulted for 3 different chemical types of GHRP, i.e., benzolactam, spiropiperidine and GHRP-6/GHS. Since each GHRP/GHS but not all active GHRPs had an essential positive charged nitrogen atom at the N-terminus, the nonfunctional mutated receptor was explained by eliminating the counter ion interaction between each of the GHRPs and the receptor. The transmembrane domain 2, 5 and 6 mutations induced different effects on the binding of these three chemically distinct GHRPs. This led to the speculation that these three GHRPs probably bind to the receptor site by different orientations.

The impact of the identification of the ghrelin receptor has had a major role in further understanding the potential significance and dimensions of the new ghrelin system. Salient examples are those which support that GHRPs and ghrelin regulate both GH secretion and food intake via the same receptor. Most complementary and biologically meaningful were the findings of the colocalization of the ghrelin/GHS-R on the GHRH and neuropeptide Y (NPY) neurons in the hypothalamic arcuate nucleus and that GH excess and deficiency decreased and increased the arcuate ghrelin receptors, respectively. Additionally, it was demonstrated that the somatotrophs are the pituitary cell site of the receptor for ghrelin/GHRPs and that the receptors are increased by GHRH, glucocorticoids and thyroid hormone.

The Central Nervous System and peripheral distribution of mRNA for the GHRP/GHS-R in normal tissue as described by Guan and coworkers (10) has been extended to endocrine and non-endocrine tumors but the exact distribution, type and function has yet to be established. Consistently striking has been the marked differences in chemistry between the GHRPs and ghrelin with the same relative binding affinity for the natural ghrelin receptor. Furthermore, the degree of binding activity and GH releasing activity parallel each other. The nonparallel exceptions of these and other biological activities among the GHRPs and between the GHRPs and ghrelin have been infrequent but nevertheless notable because of the possibility that they may be representative of receptor subtypes or new and unexpected actions of GHRPs and/or ghrelin that are peptide dependent (11). A number of potentially valuable physiological and pharmacological actions may evolve from the results of these preliminary binding and/or new biological actions. Cloning a putative ghrelin/GHS subtype receptor would be of major significance.

4.   BIOLOGICAL ACTIONS

To select the biological actions that are considered functional historical milestones is somewhat dependent on what one conceptually considers the primary basic physiological role of ghrelin to be. Obviously there is and will continue to be a difference of opinion as new data is reported. It is my opinion currently that ghrelin is mainly to attenuate effects of starvation while during normal food intake the physiological role of the ghrelin system is relatively minor. In this conceptual model, plasma ghrelin, GH levels, hunger and food intake coordinately would be increased during starvation and during satiation, they would be relatively decreased. The degree or duration of starvation required to activate the ghrelin system physiologically may be subject dependent as well as modulation dependent by specific metabolic factors and hormones.

Although the functional historic milestones, in comparison to the chemical and receptor milestones, have been more difficult to select, the physiological regulation of GH and food intake are considered important. Increased GH and food intake have been well established for several GHRPs and also, more recently, for ghrelin in animals and humans. The finding that GHRPs and ghrelin have the same biological actions has allowed the direct comparison and confirmation of these results. An important example is that neuroendocrine data demonstrating the relevant actions in the regulation of GH secretion and food intake on the hypothalamus and pituitary have been published for both GHRP and ghrelin. Furthermore, the basic and clinical results with GHRPs and now ghrelin have been meaningfully complementary (12). The general implication from these findings is that they become a fundamental basis for concluding that ghrelin is an important new biological system and thus a major future challenge will be to define its physiological role. To what degree the ghrelin system may have a non-endocrine role in addition to an endocrine role is now being investigated.

Since the isolation of GHRH in 1982, comparative studies with GHRP-6 over the following three years revealed not only the differences but also the complementary action of these two peptides on GH release. In 1984, we hypothesized that the GH releasing activity of GHRP reflected the activity of another endogenous factor or hypophysiotropic hormone that regulated GH release. A direct pituitary and more gradually a hypothalamic action of GHRPs on releasing GH was established. The neuroendocrine pioneering studies of S. Dickson and I. Robinson (1993) clearly demonstrated a direct activation of hypothalamic arcuate neurons by GHRP. These results have been extended with both GHRPs and ghrelin to include activation of a subpopulation of GHRH but not somatostatin neurons and more recently demonstration of GHRP/ghrelin receptors on the GHRH arcuate neurons. As stated below, GHRH is essential to the *in vivo* GH releasing action of GHRP and ghrelin but what is less clear is the degree that the GHRP/ghrelin GH releasing action *in vivo* is the result of releasing GHRH. These details remain unanswered; however, fundamental evidence does enhance the GHRH GH releasing action and likewise GHRH enhances the GHRP/ghrelin GH releasing action. This connection does represent an effective and highly versatile biological regulatory system.

As predicted, the first study in 1989 in normal humans revealed that GHRP (GHRP-6) released large amounts of GH. Also, GHRP-6+GHRH released GH synergistically. Importantly, it was reported, by Takaya and coworkers in 2000 (13) and by Hataya and coworkers in 2001 (14), that ghrelin and ghrelin+GHRH induced the same dramatic effect on GH release. Both GHRP-6 and ghrelin slightly raised plasma prolactin, cortisol and adrenocorticotropic hormone levels but not above the normal range and this was the result of pharmacological rather than physiological actions of these peptides (15,16).

The GHRP and GHRH interaction and now the ghrelin and GHRH interaction are perceived as an important functional milestone. This has resulted from the studies of many talented investigators who have demonstrated, defined and extended the physiological and pathophysiological functional importance of this interaction. From the physiologic viewpoint, GH releasing results of GHRPs and ghrelin in animals and humans indirectly support the conceptual model of the GHRH and ghrelin interactions to complementarily release GH with GHRH being the primary determinant. Without GHRH secretion or action the *in vivo* GH releasing action of GHRP or ghrelin is markedly decreased. When ghrelin secretion is enhanced, for example in starvation, ghrelin acts globally on the hypothalamic-pituitary unit to increase GHRH release and the action of GHRH on GH release and possibly synthesis. Indirect neuroendocrine basic and biochemical findings supports that GHRPs and presumably ghrelin enhance pituitary GH synthesis, i.e., activation of arcuate GHRH neurons by GHRPs and ghrelin (17,18), localization of GHRP/GHS-R on arcuate GHRH neurons (19) and sustained enhancement of GH secretion in humans over a 30-60 day period by continuous GHRP administration. From the pathophysiological viewpoint, possible functional abnormalities of the ghrelin and GHRH interaction may occur which would result in decreased secretion of GH, i.e., deficiency of ghrelin, deficiency of GHRH and varying degrees of deficiency of both GHRH and ghrelin. Particularly noteworthy is that decreased secretion of GHRH would induce a secondary functional deficiency of ghrelin even when plasma ghrelin levels are within normal range. From the interaction viewpoint, evaluation of the relationship between GHRH and ghrelin becomes especially clinically relevant both diagnostically and therapeutically.

Enhancement of the normal pulsatile GH secretion via continuous iv administration of GHRP-6 by Thorner, Barkan, Jaffe and their coworkers in 1993 (20,21) has been selected as a functional historical milestone since it forecast GHRP/ghrelin may become a valuable new therapeutic physiological approach for increasing the physiological secretion of GH in humans. The in-depth studies in critically ill patients on continuous infusion of GHRP-2 by van den Berghe and coworkers in 1998 (22) demonstrate the normal pulsatile GH secretion and other valuable clinical and biochemical GHRP effects as well as GHRP actions in severely ill patients. Continuous subcutaneous infusion of GHRP now has been extended to 30-60 days in normal older men and women with low serum insulin-like growth factor-I levels. In these studies GHRP increased the physiological pulsatile secretion of GH and raised serum insulin-like growth factor-I (IGF-I) levels as well as IGF-binding protein-3 into the range of normal values over the entire infusion period.

The functional historical milestones certainly include the three long term GHRP studies of Laron (1997), Pihoker (1997), Mericq (1998) and their coworkers (23-25) for in each study the height of short-statured children with various degrees of GH deficiency was increased. Intranasal hexarelin

(23) or GHRP-2 (24) was administered for 6-24 months while Mericq administered single daily GHRP-2 subcutaneously for 8 months (25). Although the increase in height velocity was not marked, it was accelerated in each study and the GHRP approach is still yet to be optimized in children. In particular, these studies demonstrate chronic GHRP therapy increases anabolism in humans. Since GHRP augments the physiological secretion of GH and effectively activates the entire GH/IGF-I axis, new more optimized GHRP approaches will be significantly important. In 1998 a 2-month well-controlled study of obese men by Svensson and coworkers demonstrated that MK-0677 once daily increased lean body mass while total body fat was unchanged (26). In 2002, the detailed clinical studies of Veldhuis and coworkers revealed that estrogen regulates GH secretion in postmenopausal women by attenuating the potency of the inhibitory action of somatostatin, augmenting pituitary sensitivity to GHRH, facilitating GHRP-2 maximal GH secretion and muting the GH autonegative feedback on GHRP-2 (27).

Over a number of years, GHRPs have been utilized in the diagnosis of GH deficiency as well as other hormonal disorders such as Cushing's syndrome/disease. Impressive are the results of the combined GHRP+GHRH or GHRP alone as a diagnostic test which merits historical milestone status. These consist of GHRP-6+GHRH, hexarelin+GHRH and GHRP-2 at a high dosage to establish new dimensions in the pathophysiology of various disorders and to elucidate new insight into the physiological regulation of GH secretion and various hormones such as estrogen, testosterone, and cortisol which influence GH regulation. The in-depth high quality of these new GHRP-6, hexarelin and GHRP-2 diagnostic approaches underscore the importance of basic and clinical GHRP/ghrelin studies that have been performed by Casanueva, Dieguez and coworkers in Spain, Ghigo and coworkers in Italy and Chihara and coworkers in Japan. Also to be acknowledged is Deghenghi, who has developed several GHRPs, including hexarelin, and collaborated on many GHRP/ghrelin basic and clinical studies.

The results of the initial indirect and increasingly direct hypothalamic studies became a strong basis for convincingly believing that GHRP and eventually ghrelin would be physiologically involved in the regulation of food intake. A series of neuroendocrine results on the hypothalamic arcuate nucleus related to food intake have been obtained by Dickson, Robinson, Kamegai, Nakazato, Willesen, Wren and their coworkers (1993-2002). Some of the studies include the following: a) stimulation of NPY neurons and food intake by intra-cerebroventricular (icv), intra-peritoneal and intra-arcuate nucleus administration of GHRPs and/or ghrelin to rats; b) colocalization of GHRP/ghrelin receptor on NPY neurons; c) increase of mRNA of both orexigenic peptides, NPY and agouti-related protein, by icv ghrelin; d) decreased food intake by icv DLys³-GHRP-6 antagonist. In contrast to the *in vivo* GH releasing action of GHRP/ghrelin, the effect on stimulation of food intake in GH deficient mice and rats indicates GH but not food intake is GHRH dependent. Also, an adipogenic action of GHRP/ghrelin has been observed in GH deficient rats and mice. Understanding the implications of the differences in GHRH dependency and the adipogenic action of ghrelin is projected to be relevant physiologically and pathophysiologically.

From the hormonal-metabolic viewpoint, integrating and balancing the dual action of GHRP/ghrelin on GH and food intake is compelling (28,29). Conceptually, Tschop, Smiley and Heiman in 2000 emphasized the basic and biological importance of the coordinated regulation of GH and food intake which does characterize the uniqueness of ghrelin as a hormone with a new action and physiological role (30). This hormonal uniqueness is strongly supported by the stimulatory action of both ghrelin and GHRP on food intake as well as the most unexpected new finding that ghrelin anatomically originates primarily from the stomach. This single anatomical finding in combination with the lack of anatomical evidence that ghrelin is synthesized in the hypothalamus has had a major impact and reorientation on how the physiological role of ghrelin is envisioned. Recently preliminary evidence indicates ghrelin is synthesized at a select site in the arcuate nucleus but only in such low amounts that a dual hypothalamic-pituitary action on GH release would be unexpected and the relationship to food intake is an unknown. Notable is the increase of food intake by ghrelin in humans after peripheral administration of ghrelin (31).

The relationship between the peripheral plasma ghrelin levels in animals and humans are interrelated with normal and abnormal nutritional states. So far the limitation of this most important objective concerns the validation of the assay specifically in terms of the bioactive octanoylated ghrelin molecule. Plasma concentration of bioactive ghrelin and the relevance of this concentration on GH release and increased food intake, which may or may not be in parallel, will be equally important. The action of ghrelin on food intake, unlike that on GH secretion, does not appear to be dependent on GHRH.

5.   CONCLUSION

The selection of milestone is a subjective. The degree, context and timing of discoveries are important, and negative results are as dramatic as positive results. New discoveries are exciting; their status as milestones is yet to be determined as the science of ghrelin unfolds.

ACKNOWLEDGEMENTS

A number of innovative and exciting, potential historical milestones have not been included in this brief chapter. The primary reasons for not including them has been the page limitation, incomplete nature of the studies and/or the lack of confirmation especially those studies concerning complex biological actions or isolated observations in terms of fundamental importance. Nevertheless, to acknowledge these many valuable contributions, especially those in the category of non-endocrine actions, several GHRP/ghrelin reviews have been added to the references to include the many unnamed authors and contributions.

This work was supported in part by the National Institute of Health Grant RR05096.

REFERENCES

1. Bowers CY. Unnatural growth hormone-releasing peptide begets natural ghrelin. J Clin Endocrinol Metabol. 2001; 86:1464-69.
2. Bowers CY. GH releasing peptides (GHRPs). In: Kostyo JL and Goodman HM, eds. Handbook of Physiology. Section 7, volume V. New York: Oxford University Press, 1999; 267-97.
3. Bowers CY, Chang J, Momany F, Folkers K. Effect of the enkephalins and enkephalin analogs on the release of pituitary hormones *in vitro*. In: MacIntyre I and Szelke M, eds. Molecular Endocrinology. Amsterdam: Elsevier/North-Holland Biomedical Press, 1977; 287-92.
4. Kojima M, Hosoda H, Date Y, et al. Ghrelin is a growth hormone releasing acetylated peptide from the stomach. Nature. 1999; 402:656-60.
5. Hughes J, Smith TW, Kosterlitz HW, Fothergill LA, Morgan BA, Morris HR. Identification of two related pentapeptides from the brain with potent opiate agonist activity. Nature. 1975; 258:577-80.
6. Kojima M, Hosoda H, Matsuo H, Kangawa K. Ghrelin: discovery of the natural endogenous ligand for the growth hormone secretagogue receptor. Trends Endocrinol Metabol. 2001; 12:118-22.
7. Howard AD, Feighner SD, Cully DF, et al. A receptor in pituitary and hypothalamus that functions in growth hormone release. Science 1996; 273:974-77.
8. Smith RG, Van der Ploeg LHT, Howard AD, et al. Peptidomimetic regulation of growth hormone secretion. Endocrine Rev. 1997; 18:621-45.
9. Feighner SD, Howard AD, Prendergast K, et al. Structural requirements for the activation of the human growth hormone secretagogue receptor by peptide and nonpeptide secretagogues. Mol Endocrinol. 1998; 12:137-45.
10. Guan XM, Yu H, Palyha OC, et al. Distribution of mRNA encoding the growth hormone secretagogue receptor in brain and peripheral tissues. Brain Res Mol Brain Res. 1997; 48:23-9.

11. Muccioli G, Tschop M, Papotti M, Deghenghi R, Heiman M, Ghigo E. Neuroendocrine and peripheral activities of ghrelin: implications in metabolism and obesity. Eur J Pharmacol. 2002; 440:235-54.

12. Dieguez C and Casanueva FF. Ghrelin: a step forward in the understanding of somatotroph cell function and growth regulation. Eur J Endocrinol. 2000; 142:413-17.

13. Takaya K, Ariyasu H, Kanamoto N, et al. Ghrelin strongly stimulates growth hormone release in humans. J Clin Endocrinol Metab. 2000; 85:4908-11.

14. Hataya Y, Akamizu T, Takaya K, et al. A low dose of ghrelin stimulates growth hormone (GH) release synergistically with GH-releasing hormone in humans. J Clin Endocrinol Metab. 2001; 86:4552-5.

15. Ghigo E, Arvat E, Muccioli G, Camanni F. Growth hormone releasing peptides. Eur J Endocrinol. 1997; 136:445-60.

16. Bowers CY. Growth hormone-releasing peptide (GHRP). Cell Molec Life Sci. 1998; 54:1316-29.

17. Dickson SL, Luckman SM. Induction of c-fos messenger ribonucleic acid in neuropeptide Y and growth hormone (GH)-releasing factor neurons in the rat arcuate nucleus following systemic injection of the GH secretagogue, GH-releasing peptide-6. Endocrinology. 1997; 138:771-7.

18. Kamegai J, Tamura H, Shimizu T, Ishii S, Sugihara H, Wakabayashi I. Chronic central infusion of ghrelin increases hypothalamic neuropeptide Y and Agouti-related protein mRNA levels and body weight in rats. Diabetes. 2001; 50:2438-43.

19. Tannenbaum GS, Lapointe M, Beaudet A, Howard AD. Expression of growth hormone secretagogue-receptors by growth hormone-releasing hormone neurons in the mediobasal hypothalamus. Endocrinology. 1998; 139:4420-3.

20. Hartman ML, Veldhuis JD, Thorner MO. Normal control of growth hormone secretion. Horm Res. 1993; 40:37-47.

21. Jaffe CA, Ho PJ, Demott-Friberg R, Bowers CY, Barkan AL. Effects of a prolonged growth hormone (GH)-releasing peptide infusion on pulsatile GH secretion in normal men. J Clin Endocrinol Metab. 1993; 77:1641-7.

22. Van den Berghe G, de Zegher F, Baxter RC, et al. Neuroendocrinology of prolonged critical illness: effects of exogenous thyrotropin-releasing hormone and its combination with growth hormone secretagogues. J Clin Endocrinol Metab. 1998; 83:309-19.

23. Laron Z, Frenkel J, Deghenghi R, Anin S, Klinger B, Silbergeld A. Intranasal administration of the GHRP hexarelin accelerates growth in short children. Clin Endocrinol. 1995; 43:631-5.

24. Pihoker C, Badger TM, Reynolds GA, Bowers CY. Treatment effects of intranasal growth hormone releasing peptide-2 in children with short stature. J Endocrinol. 1997; 155:79-86.

25. Mericq V, Cassorla F, Salazar T, et al. Effects of eight months treatment with graded doses of a growth hormone (GH)-releasing peptide in GH-deficient children. J Clin Endocrinol Metab. 1998; 83:2355-60.

26. Svensson J, Lonn L, Jansson JO, et al. Two-month treatment of obese subjects with the oral growth hormone (GH) secretagogue MK-677 increases GH secretion, fat-free mass, and energy expenditure. J Clin Endocrinol Metab. 1998; 83:362-9.

27. Veldhuis JD, Evans WS, Anderson SM, Bowers CY. Sex-steroid hormone modulation of the tripeptidyl control of the human somatropic axis. J Anti-Aging Med. 2002; 5:81-111.

28. Inui A. Ghrelin: an orexigenic and somatotrophic signal from the stomach. Nat Rev Neurosci. 2001; 2:551-60.

29. Horvath TL, Diano S, Sotonyi P, Heiman M, Tschop M. Minireview: ghrelin and regulation of energy balance- a hypothalamic perspective. Endocrinology 2001; 142:4163-9.

30. Tschop M, Smiley DL, Heiman ML. Ghrelin induces adiposity in rodents. Nature. 2000; 407:908-13.

31. Wren AM, Seal LJ, Cohen MA, et al. Ghrelin enhances appetite and increases food intake in humans. J Clin Endocrinol Metab. 2001; 86:5992-5

Chapter 2

DISCOVERY OF GHRELIN, AN ENDOGENOUS LIGAND FOR THE GROWTH-HORMONE SECRETAGOGUE RECEPTOR

Masayasu Kojima[1,*], Hiroshi Hosoda[2,3], Hiroyuki Kaiya[2] & Kenji Kangawa[2,3]
[1]Molecular Genetics, Institute of Life Science, Kurume University, Kurume, Fukuoka 839-0861, Japan; [2]Department of Biochemistry, National Cardiovascular Center Research Institute, 5-7-1 Fujishirodai, Suita, Osaka 565-8565, Japan; [3]Translational Research Center, Kyoto University Hospital, 54 Shogoin Kawahara-cho, Sakyo-ku, Kyoto 606-8507, Japan.
**e-mail:* *mkojima@lsi.kurume-u.ac.jp*

Abstract: Small synthetic molecules called growth-hormone secretagogues (GHS) stimulate the release of growth hormone (GH) from the pituitary. They act through the GHS-receptor (GHS-R), a G-protein-coupled receptor. We purified and identified the endogenous ligand for the GHS-R and named it "ghrelin," after a word root ("ghre") in Proto-Indo-European languages meaning "grow". Ghrelin is a peptide hormone in which the third amino acid, usually a serine but in some species a threonine, is modified by a fatty acid; this modification is essential for ghrelin's activity. Ghrelin is an essential hormone for maintaining GH release and energy homeostasis in vertebrates.

Key words: ghrelin, growth hormone secretagogue, acylation, stomach

1.   INTRODUCTION

Growth hormone secretagogues (GHS) are small synthetic molecules that stimulate growth hormone (GH) release from the pituitary. The first GHS, reported in 1976 by C.Y. Bowers, was a Met-enkephalin analog and had weak GH-releasing activity (1,2). Since then, peptidergic and peptidomimetic compounds with potent GH-releasing activity have been developed (3,4). GHS act through a specific G-protein-coupled receptor

(GPCR), named GHS-receptor (GHS-R), which is distinct from the receptor that binds GH-releasing hormone (GHRH). This receptor was for some time an example of an "orphan GPCR", that is a GPCR with no known natural ligand. Employing the "reverse pharmacology" paradigm with a stable cell line expressing GHS-R, we searched for the endogenous ligand of GHS-R.

2.   GHS RECEPTOR SUPERFAMILY

Figure 2-1 shows a dendrogram of GHS-R and its related receptor superfamily.

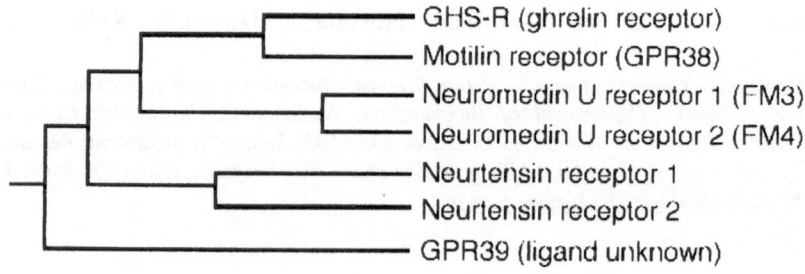

GHS-R (ghrelin receptor)
Motilin receptor (GPR38)
Neuromedin U receptor 1 (FM3)
Neuromedin U receptor 2 (FM4)
Neurtensin receptor 1
Neurtensin receptor 2
GPR39 (ligand unknown)

*Figure 2-1.*Dendrogram alignment of the GHS-R and other GPCR.

In 1996, the GHS-R was identified by expression cloning and showed to be a typical G-protein-coupled seven-transmembrane receptor (5). The receptor most closely related to GHS-R is the motilin receptor (GPR38) (6,7); the human forms of these receptors share 52% amino acid homology. Two receptors for neuromedin U (NMU-R1 and -R2), a neuropeptide that promotes smooth muscle contraction and suppresses food intake, are also homologous to GHS-R (8-11). Because motilin and NMU are found mainly in gastrointestinal organs, it was speculated that the endogenous ligand for GHS-R may be another gastrointestinal peptide. This speculation was confirmed finally by the isolation of ghrelin from stomach tissue (12).

3.   PURIFICATION AND IDENTIFICATION OF RAT GHRELIN

Classical methods for discovering unknown peptide hormones are based on assays that monitor the physiological effects induced by the hormones. After a novel hormone is identified, the structure of its receptor is then determined. Recent progress in genomic analysis, however, has reversed the process of hormone discovery: first, receptors are identified, and then the structure of its activating hormone and the physiological function of the system are determined. This powerful method, known as "reverse pharmacology" or the "orphan receptor" strategy, has provided a new tool for hormone research (13-16). Classical methods were often limited in that their assay systems were not specific for ligand-receptor complexes, whereas this new method has employed in particular the measurement of second-messenger or electrophysiological changes in receptor-expressing cells or *Xenopus* oocytes. GHS-R was known to bind artificial GHS such as GH-releasing-peptide-6 or hexarelin, providing a convenient positive control for an assay used to identify the endogenous ligand. A cultured cell line expressing GHS-R was established and used to identify tissue extracts that could stimulate the receptor, as monitored by increases in $[Ca^{2+}]_i$ levels (5,17).

The purification of ghrelin is described in Figure 2-2 (12).

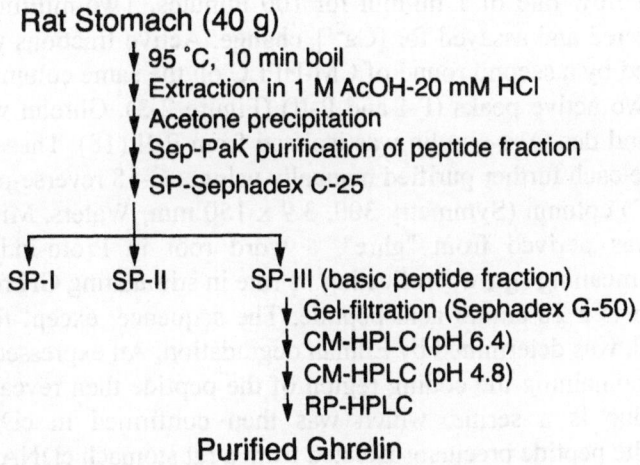

Figure 2-2. Purification of ghrelin from rat stomach tissue. AcOH: acetic acid; CM: carboxymethyl; RP-HPLC: reverse-phase high-performance chromatography.

Fresh rat stomach tissues (40 g) were minced and boiled for 10 minutes in 5x volumes of water to inactivate intrinsic proteases. The solution was adjusted to 1 M acetic acid (AcOH)-20 mM HCl. Boiled stomach tissue was homogenized with a Polytron mixer. The supernatant of the extracts was concentrated to approximately 40 ml by evaporation. A 2x volume of acetone was added to the concentrate for acetone-precipitation in 66% acetone. Acetone was removed from the isolated supernatants by evaporation. The supernatant was loaded onto a 10-g Sep-Pak C18 column (Waters, Milford, MA), pre-equilibrated in 0.1% trifluoroacetic acid (TFA). The Sep-Pak cartridge was washed with 10% acetonitrile $(CH_3CN)/0.1\%$ TFA, and the peptide fraction was eluted in 60% $CH_3CN/0.1\%$ TFA. The eluate was evaporated and lyophilized. The residual materials were redissolved in 1 M AcOH and adsorbed on a SP-Sephadex C-25 column (H^+-form, Pharmacia, Uppsala, Sweden), pre-equilibrated with 1 M AcOH. Successive elutions with 1 M AcOH, 2 M pyridine and 2 M pyridine-AcOH (pH 5.0) yielded three fractions- SP-I, SP-II and SP-III (basic peptide fraction). The SP-III fraction was lyophilized (dry weight: 39 mg) and applied to a Sephadex G-50 gel-filtration column (1.8 x 130 cm) (Pharmacia, Uppsala, Sweden). Five-milliliter fractions were collected, and a portion of each fraction was subjected to the $[Ca^{2+}]_i$-change assay using CHO-GHSR62 cells. Active fractions were separated by carboxymethyl (CM)-ion exchange high performance liquid chromatography (HPLC) on a TSK CM-2SW column (4.6 x 250 mm, Tosoh, Tokyo, Japan) using an ammonium acetate gradient $(HCOONH_4)$ (pH 6.4) of 10 mM to 1 M in the presence of 10% CH_3CN at flow rate of 1 ml/min for 100 minutes. Two-milliliter fractions were collected and assayed for $[Ca^{2+}]_i$ change. Active fractions were further fractionated by a second round of CM-HPLC on the same column at pH 4.8, yielding two active peaks (P-I and P-II) (Figure 2-3). Ghrelin was purified from P-I and des-Q14-ghrelin was isolated from P-II (18). These two active peaks were each further purified manually using a C-18 reverse-phase HPLC (RP-HPLC) column (Symmetry 300, 3.9 x 150 mm, Waters, Milford, MA). Ghrelin was derived from "ghre", a word root in Proto-Indo-European languages meaning "grow," reflecting its role in stimulating GH release.

Ghrelin is a 28-amino acid peptide. The sequence, except for the third amino acid, was determined by Edman degradation. An expressed-sequence-tag clone containing the coding region of the peptide then revealed that the third residue is a serine, which was then confirmed in cDNA clones encoding the peptide precursor isolated from a rat stomach cDNA library.

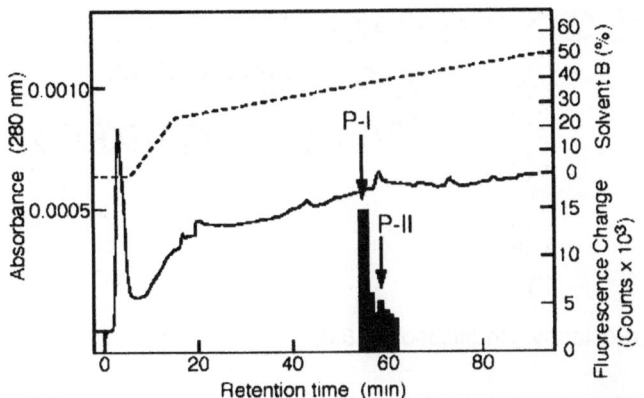

Figure 2-3. CM-ion-exchange HPLC (pH 4.8) of rat stomach extracts. Ghrelin and des-Gln[14]-ghrelin were purified from P-I and P-II, respectively.

A 28-amino acid peptide based on the third serine-containing cDNA sequence was synthesized and its characteristics compared with those of purified ghrelin. This comparison revealed the following: a) the synthetic peptide, unlike the purified peptide, did not increase $[Ca^{2+}]_i$ levels in GHS-R-expressing cells; b) the retention time of purified ghrelin in RP-HPLC was longer than that of the synthetic peptide; and c) by mass spectrometric analysis, the molecular weight (M_r) of purified ghrelin (M_r: 3,315) was 126 daltons greater than that of the synthetic peptide (M_r: 3,189).

The most probable modification was that of a fatty acid, n-octanoic acid. When the hydroxyl group of serine 3 of the synthetic peptide (M_r: 3,189) was esterified by n-octanoic acid (M_r: 144), the resulting modified peptide was of the same molecular weight as purified ghrelin (M_r: 3,315). Moreover, the acyl modification appeared to increase the hydrophobicity of the peptide, explaining the increase in retention time when it was subjected to HPLC.

The peptide was then synthesized with a serine 3 n-octanoyl acid modification and its characteristics were compared with those of purified ghrelin. This synthetic modified peptide comigrated with purified ghrelin on RP-HPLC, and mass spectrometric fragmentation patterns of the synthetic modified peptide and purified ghrelin were the same. Moreover, the synthetic modified peptide had the same effect as purified ghrelin on the GHS-R-expressing cells. These results confirmed the primary structure of ghrelin, an octanoyl-modified peptide (Figure 2-4).

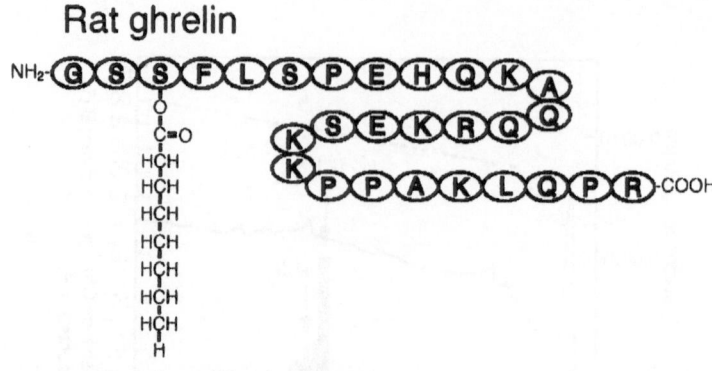

n-octanoyl modification (C8:0)

Figure 2-4. Structure of rat ghrelin. The third amino acid, serine, is modified by an acyl acid, n-octanoic acid, and this modification is essential for ghrelin's activity.

4.   MINOR MOLECULAR FORMS OF HUMAN GHRELIN

The major form of active human ghrelin in the stomach is, like rat ghrelin, a 28-amino acid peptide with an octanoyl modification at its third amino acid, serine. During the course of purification, we isolated several minor forms of ghrelin peptides (19). These peptides were classified into four groups by the type of acylation observed at serine 3: non-acylated, octanoylated (C8:0), decanoylated (C10:0), and possibly decenoylated (C10:1). All peptides found are either 27 or 28 amino acids in length, the former lacking the C-terminal arginine 28, and are derived from the same ghrelin precursor through two alternative pathways. Synthetic octanoylated and decanoylated ghrelins induce $[Ca^{2+}]_i$ increases in GHS-R-expressing cells and stimulate GH release in rats to similar degrees.

5.   SEQUENCE COMPARISON OF MAMMALIAN GHRELIN

We also determined the amino-acid sequences of mouse, bovine, porcine, ovine and canine ghrelins (Figure 2-5) (20) and found them to be well conserved.

```
                    *
Human:    GSSFLSPEHQRVQQRKESKKPPAKLQPR
Rat:      GSSFLSPEHQKAQQRKESKKPPAKLQPR
Mouse:    GSSFLSPEHQKAQQRKESKKPPAKLQPR
Bovine:   GSSFLSPEHQELQ-RKEAKKPSGRLKPR
Porcine:  GSSFLSPEHQKVQQRKESKKPAAKLKPR
Ovine:    GSSFLSPEHQKLQ-RKEPKKPSGRLKPR
Canine:   GSSFLSPEHQKLQQRKESKKPPAKLQPR
```

Figure 2-5. Alignment of amino-acid sequences of mammalian ghrelins. Set of identical residues are shaded. The asterisk indicates the third amino acid, serine, which undergoes an acyl modification.

In particular, the N-terminal 10 amino acids are universally identical. The structural conservation and strictly required modification indicate that the N-terminal portion of the peptide is of central importance in the activity of the peptide. Bovine and ovine ghrelins comprise 27 amino acids, like des-Gln^{14}-ghrelin, the second endogenous form of rat ghrelin produced through an alternative splicing of ghrelin mRNA.

6.   GHRELIN PRECURSOR AND DES-GLN^{14}-GHRELIN

The rat and human ghrelin precursors are both composed of 117 amino acids (12). In these precursors, the ghrelin sequence immediately follows that of the signal peptide. At the C-terminus of the ghrelin sequence, processing occurs at an uncommon proline-arginine recognition site. In the rat stomach, two isoforms of mRNA encoding pro-ghrelin are produced from the gene by an alternative splicing mechanism. One mRNA encodes the ghrelin precursor, and another encodes a precursor for des-Gln^{14}-ghrelin, a peptide identical to ghrelin, but with a deletion of glycine 14 (18). This deletion results from the use of the CAG codon, which encodes glutamine 14 as well as acts as a splicing signal. Thus, two types of active ghrelin peptide are produced in rat stomach, ghrelin and des-Gln^{14}-ghrelin (Figure 2-3). However, des-Gln^{14}-ghrelin is only present in low amounts in the stomach and full-length ghrelin is the major active form.

7.   NON-MAMMALIAN GHRELIN

7.1   Bullfrog ghrelin

Amphibian ghrelin has been identified from the stomach of bullfrog (Figure 2-6) (21).

```
                         *
Human:       GSSFLSP-EHQRVQQRKESKKPPAKLQPR
Rat:         GSSFLSP-EHQKAQQRKESKKPPAKLQPR
Bullfrog:    GLTFLSPADMQKIAERQSQNKLRHGNMN
Chicken:     GSSFLSP-TYKNIQQQKDTRKPTARLH
Goldfish:    GTSFLSPA--QKPQGRRPPRM
Eel:         GSSFLSP-S-QRPQG-KD-KKPPRV-NH2
```

Figure 2-6. Alignment of amino-acid sequences of human, rat, bullfrog, chicken, goldfish, and eel ghrelins. Sets of residues identical in at least three species are shaded. The asterisk indicates the acylated third amino acid, serine or threonine. Eel, and possibly goldfish, ghrelin has an amide structure at its C-terminus.

The three forms of ghrelin identified in bullfrog stomach, each comprising 27 or 28 amino acids, possess 29% sequence identity with human ghrelin. A unique threonine at the third amino acid position in bullfrog ghrelin differs from the serine present in the mammalian ghrelins; this threonine 3 is acylated by either n-octanoic or n-decanoic acid. The three ghrelin isoforms in frog have the following sequences: ghrelin-28, GLT (O-n-octanoyl)FLSPADMQKIAERQSQNKLRHGNMN; ghrelin-27, GLT(O-n-octanoyl)FLSPADMQKIAERQSQNKLRHGNM; and ghrelin-27-C10, GLT(O-n-decanoyl)FLSPADMQKIAERQSQNKLRHGNM. Northern blot analysis demonstrated that ghrelin mRNA is expressed predominantly in the stomach. Low levels of gene expression were observed in the heart, lungs, small intestine, gall bladder, pancreas, and testes, as revealed by reverse-transcription-polymerase-chain-reaction (RT-PCR) analysis. Bullfrog ghrelin stimulated the secretion of both GH and prolactin in dispersed bullfrog pituitary cells with a potency 2-3 orders of magnitude greater than that of rat ghrelin. Bullfrog ghrelin, however, was only minimally effective in elevating plasma GH levels following intravenous injection into rats. These results indicate that although the mechanism by which ghrelin regulates GH secretion is evolutionarily conserved, the structural differences in the various ghrelins result in species-specific receptor binding properties.

7.2   Chicken ghrelin

Chicken (*Gallus gallus*) ghrelin is composed of 26 amino acids (GSSFLSPTYKNIQQQKDTRKPTARLH) and possesses 54% sequence identity with human ghrelin (Figure 2-6) (22). The serine residue at position 3 is conserved between the chicken and mammalian homologs, as is its acylation by either n-octanoic or n-decanoic acid. Chicken ghrelin mRNA is expressed predominantly in the stomach, where it is present in the proventriculus but absent in the gizzard. Using reverse transcriptase-polymerase chain reaction analysis, low levels of expression were also detectable in brain, lung, and intestine. Administration of chicken ghrelin increases plasma GH levels in both rats and chicks, with a potency similar to that of rat or human ghrelin. In addition, chicken ghrelin increases plasma corticosterone levels in growing chicks at a lower dose than in mammals.

7.3   Fish ghrelin

The structure of goldfish ghrelin was determined by cDNA analysis (Figure 2-6) (23). The 490-bp cDNA encodes a 103-amino-acid preproghrelin which has a 26-amino-acid signal region, possibly a 19-amino-acid mature peptide and a 55-amino-acid C-terminal peptide region (Figure 2-6). Goldfish ghrelin may have a C-terminal amide structure (methionine-amide), since the C-terminal processing site is GRR, a typical signal for peptide C-terminal amidation. Moreover, the mature goldfish ghrelin peptide has a putative cleavage site and an amidating signal (GRR) within its sequence. Endogenous forms of goldfish ghrelin, thus, may be two amidated peptides.

We purified ghrelin from stomach extracts of a teleost fish, the Japanese eel (*Anguilla japonica*), and found that it contains an amide structure at its C-terminal end (Figure 2-6) (24). Two forms of ghrelin, each comprising 21 amino acids, were identified: eel ghrelin-21, with an n-octanoyl modification, and eel-ghrelin-21-C10, with an n-decanoyl modification.

8.   CONCLUSION

The discovery of the endogenous ligand for GHS-R, ghrelin, opens up a new field in hormone research. Ghrelin is a peptide hormone, in which the third amino acid, usually serine but in some species threonine, is modified by an acyl acid; this modification is essential for ghrelin's activity. Ghrelin exists not only in mammalian species, but also in non-mammalian species such as frog, chicken, and fish. Ghrelin, thus, may be an essential hormone for maintaining GH release and energy homeostasis in vertebrates. The acyl-modified structure of ghrelin reveals an unknown pathway for propeptide processing. There remain many interesting questions regarding ghrelin-related biology: these include the identification of the pathways regulating ghrelin release in the stomach and the mechanism of the n-octanoyl modification. Further research will answer these questions and elucidate the biochemical and physiological roles of this novel hormone.

REFERENCES

1. Bowers CY, Momany F, Reynolds GA, Chang D, Hong A, Chang K. Structure-activity relationships of a synthetic pentapeptide that specifically releases growth hormone *in vitro*. Endocrinology. 1980; 106:663-7.
2. Bowers CY, Momany FA, Reynolds GA, Hong A. On the *in vitro* and *in vivo* activity of a new synthetic hexapeptide that acts on the pituitary to specifically release growth hormone. Endocrinology. 1984; 114:1537-45.
3. Smith RG, Cheng K, Schoen WR, et al. A nonpeptidyl growth hormone secretagogue. Science. 1993; 260:1640-3.
4. Smith RG, Palyha OC, Feighner SD, et al. Growth hormone releasing substances: types and their receptors. Horm Res. 1999; 51 Suppl 3:1-8.
5. Howard AD, Feighner SD, Cully DF, et al. A receptor in pituitary and hypothalamus that functions in growth hormone release. Science. 1996; 273:974-7.
6. Feighner SD, Tan CP, McKee KK, et al. Receptor for motilin identified in the human gastrointestinal system. Science. 1999; 284:2184-8.
7. McKee KK, Tan CP, Palyha OC, et al. Cloning and characterization of two human G protein-coupled receptor genes (GPR38 and GPR39) related to the growth hormone secretagogue and neurotensin receptors. Genomics. 1997; 46:426-34.
8. Fujii R, Hosoya M, Fukusumi S, et al. Identification of neuromedin U as the cognate ligand of the orphan G protein-coupled receptor FM-3. J Biol Chem. 2000; 275:21068-74.
9. Hosoya M, Moriya T, Kawamata Y, et al. Identification and functional characterization of a novel subtype of neuromedin U receptor. J Biol Chem. 2000. 275:29528-32.
10. Howard AD, Wang R, Pong SS, et al. Identification of receptors for neuromedin U and its role in feeding. Nature. 2000; 406:70-4.

11. Kojima M, Haruno R, Nakazato M, et al. Purification and identification of neuromedin U as an endogenous ligand for an orphan receptor GPR66 (FM3). Biochem Biophys Res Commun. 2000; 276:435-8.

12. Kojima M, Hosoda H, Date Y, Nakazato M, Matsuo H, Kangawa K. Ghrelin is a growth-hormone-releasing acylated peptide from stomach. Nature. 1999; 402:656-60.

13. Civelli O. Functional genomics: the search for novel neurotransmitters and neuropeptides. FEBS Lett. 1998; 430:55-8.

14. Civelli O, Nothacker HP. Functional genomics and the discovery of new drug targets. Diabetes Technol Ther. 1999; 1:71-6.

15. Civelli O, Nothacker HP, Saito Y, Wang Z, Lin SH, Reinscheid RK. Novel neurotransmitters as natural ligands of orphan G-protein-coupled receptors. Trends Neurosci. 2001; 24:230-7.

16. Howard AD, McAllister G, Feighner SD, et al. Orphan G-protein-coupled receptors and natural ligand discovery. Trends Pharmacol Sci. 2001; 22:132-40.

17. McKee KK, Palyha OC, Feighner SD, et al. Molecular analysis of rat pituitary and hypothalamic growth hormone secretagogue receptors. Mol Endocrinol. 1997; 11:415-23.

18. Hosoda H, Kojima M, Matsuo H, Kangawa K. Purification and characterization of rat des-Gln14-Ghrelin, a second endogenous ligand for the growth hormone secretagogue receptor. J Biol Chem. 2000; 275:21995-2000.

19. Hosoda H, Kojima M, Mizushima T, Shimizu S, Kangawa K. Structural divergence of human ghrelin. Identification of multiple ghrelin-derived molecules produced by post-translational processing. J Biol Chem. 2003; 278:64-70.

20. Tomasetto C, Wendling C, Rio MC, Poitras P. Identification of cDNA encoding motilin related peptide/ghrelin precursor from dog fundus. Peptides. 2001; 22:2055-9.

21. Kaiya H, Kojima M, Hosoda H, et al. Bullfrog ghrelin is modified by n-octanoic acid at its third threonine residue. J Biol Chem. 2001; 276:40441-8.

22. Kaiya H, Van Der Geyten S, Kojima M, et al. Chicken ghrelin: purification, cDNA cloning, and biological activity. Endocrinology. 2002; 143:3454-63.

23. Unniappan S, Lin X, Cervini L, et al. Goldfish ghrelin: molecular characterization of the complementary deoxyribonucleic acid, partial gene structure and evidence for its stimulatory role in food intake. Endocrinology. 2002; 143:4143-6.

24. Kaiya H, Kojima M, Hosoda H, et al. Amidated fish ghrelin: purification, cDNA cloning in the Japanese eel and its biological activity. J Endocrinol. 2003; 176:415-23.

Chapter 3

KNOWN AND UNKNOWN GROWTH HORMONE SECRETAGOGUE RECEPTORS AND THEIR LIGANDS

Giampiero Muccioli, Fabio Broglio[1], Elena Tarabra & Ezio Ghigo[1]
Departments of Pharmacology and [1]Internal Medicine, University of Turin Medical School, Turin, Italy

Abstract: The discovery of ghrelin is a typical example of reverse pharmacology: first the synthetic analogues (i.e. peptidyl and non-peptidyl growth hormone secretagogues (GHS)); second the receptor; third the natural ligand for this orphan receptor, i.e. ghrelin. Ghrelin is providing new understanding about how the gastrointestinal tract and nutritional intake regulate appetite, food intake and energy expenditure as well as the function of hypothalamus-pituitary axis, particularly the somatotroph function. However, ghrelin and its synthetic analogues also exert other peripheral, endocrine and nonendocrine actions including influence on the endocrine pancreas and glucose metabolism, on gastroenteropancreatic and cardiovascular functions, as well as modulation of cell proliferation. The classical ghrelin receptor, i.e., the GHS receptor type 1a (GHS-R1a) is widely distributed and mediates many of the biological actions of ghrelin and synthetic GHS. GHS-R type 1b a splice variant of the GHS-R1a, is even more widely distributed in central and peripheral tissues but its function is still unknown. Other GHS-R subtypes, i.e., other ghrelin receptors are likely to exist. In fact, GHS-R1a is bound by acylated ghrelin only but the existence of a receptor able to bind ghrelin independently of its acylation has been already demonstrated in agreement with evidence that unacylated ghrelin, although devoid of endocrine actions, possesses cardiovascular effects and modulates cell proliferation. Moreover, the existence of specific receptors mediating some cardiovascular actions of peptidyl GHS only and that do not recognize ghrelin has been demonstrated. Whether ghrelin is the sole ligand or one of a number of ligands activating the GHS-R1a is, at present, under investigation; on the other hand, it has also to be clarified whether the GHS-R used for ghrelin isolation is the sole receptor or one of a group of receptors for one or more than one ligand.

Key words: growth hormone secretagogues, receptors

1.   INTRODUCTION

The discovery of ghrelin, a novel growth hormone (GH)-releasing peptide (GHRP) of gastric origin, ends one mystery started about thirty years ago with the invention of non-natural GH secretagogues (GHS) and reveals questions about how the gastrointestinal tract and nutritional intake regulate appetite, food intake and energy expenditure, as well as the function of the hypothalamus-pituitary axis, particularly the somatotroph function. In fact ghrelin was discovered as a natural ligand of the orphan GHS receptor (GHS-R) type-1a (GHS-R1a) which, in turn had been shown to be specific for synthetic GHS. Thus, the discovery of ghrelin is a typical example of reverse pharmacology: first the analogues (of a hitherto unknown endogenous substance) were obtained, second the receptor was cloned and third the natural ligand for this orphan receptor, i.e. ghrelin, was isolated. In this chapter, the biochemical-pharmacological characteristics and the functional significance of ghrelin/GHS receptors are discussed.

2.   HISTORICAL BACKGROUND

2.1   From synthetic GHS to the natural ligand ghrelin via receptor identification

The pioneering work of Bowers and his collaborators (1) with the prototype hexapeptide GHRP-6 (2) established the possibility that GH release could be manipulated by small GHS derived from short peptides, such as met-enkephalin, able to liberate GH acting at the hypothalamic but not at the pituitary level. Although overshadowed by the identification of the hypothalamic GH-releasing hormone (GHRH) during the 1980s, the interest in the field of GHS was raised again some years later when it was observed that GHS operated through non-GHRH and non-somatostatinergic mechanisms (3,4) acting on receptors distinct from those of GHRH, somatostatin (SRIF) or enkephalins (5-7). The ability of GHS to synergize with GHRH and to show remarkable potency in man (8) stimulated the development of more potent peptides (hexarelin) and orally active smaller peptide (JMV-1843) and non-peptide (MK-0677) mimetics (9-11). Some years later, in 1996, the activation of phospholipase C (PLC)/inositol trisphosphate (IP_3) pathway and the rise in intracellular Ca^{2+} induced by GHS associated to a MK-0677 binding studies and a strategy of expression cloning were adopted by the group of R. Smith, at that time at Merck Sharp & Dohme Research Lab (USA), and allowed isolation in the hypothalamus-pituitary axis a cDNA encoding the GHS-R (12-14). This, called GHS-R1a, constituted the first of a new branch of a large family of receptors (see below), and its discovery implied that GHS may mimic the action of unknown endogenous hormonal factors (as long predicted by C. Bowers). The cloning of GHS-R was important because, in addition to the information *per se*, its expression in a cell line allowed identifying the unknown endogenous ligand. Using this strategy, Kojima and Kangawa in 1999 (15) isolated and chemically defined a peptide from the rat stomach that functioned as a bioactive ligand to the earlier defined receptor for GHS. This peptide was called ghrelin, a term that contains "GH" for growth hormone and "relin", a suffix for releasing substances in generic names which also represents an abbreviation for "growth-hormone-release", a characteristic effect of ghrelin (15,16). The purified ligand is made up of 28 amino acids and esterified with an octanoic acid at serine 3 residue. The acylation of the peptide regulates the extent and the direction of ghrelin transport across the blood-brain barrier (17), and it is also essential for binding the GHS-R1a (18) and for its GH-releasing activity *in vivo* (19) and for other central and peripheral endocrine and non-endocrine actions. Among these: prolactin and adrenocorticotropic hormone-releasing effects, inhibitory influence on gonadotropin and thyroid stimulating hormone secretion, orexigenic and positive effects on energy balance, anxiogenic and slow-wave sleep-promoting activity, hypotensive, cardioprotective and cardiostimulant actions, antiproliferative and immunomodulating activities, stimulation of gastric acid secretion and gut motility, adipogenic and weight gain-promoting effects, influence on placental and gonadal functions and regulation of insulin secretion and glucose metabolism (20-23). Another endogenous ligand of the GHS-R1a has also been isolated from the rat stomach and is named des-Gln[14]-ghrelin. It has the same acylation in serine 3 residue and is homologous to ghrelin except for one glutamine that is missing. It is the result of an alternative splicing of the ghrelin gene and possesses the same GH-releasing activity of ghrelin (24). Octanoic acid is not the only serine 3 modification that confers full activity to ghrelin, since multiple ghrelin-derived molecules (decanoyl and decenoyl ghrelin) produced by postranslational processing have been recently isolated from the human stomach (25). These acyl-modified ghrelin forms, that circulate in an amount far lower (octanoylated/decanoylated ratio of 3:1), stimulate GH-release in rats to a similar degree (25). A non-acylated ghrelin, which circulates in an amount far greater than the acylated form (octanoylated/des-octanolylated ratio 1:4) (26), is also biologically active. Although it is devoid of GHS-R1a binding affinity (18) and does not stimulate GH release *in vivo* (19), it is able to exert some non-endocrine actions including cardiac and antiproliferative effects, probably by binding different GHS-R subtypes common for ghrelin and des-acyl ghrelin (27-30, see also below). It is likely that further subtypes of GHS-R different from the already cloned one exist, since evident differences in the binding profile among ghrelin, and synthetic peptidyl (hexarelin) and non-peptidyl (MK-0677) have been reported (31-34). In a detailed analysis using labelled hexarelin our research group and H. Ong and colleagues in Montreal demonstrated that a wide range of non-endocrine human tissues, and in particular the heart, possess GHS binding sites specific for peptidyl GHS only (35-37). In agreement with this finding there is the fact that not all natural and synthetic GHS show the same GH-independent extra-endocrine activities (38,39). All together these studies indicate that the cloned GHS-R is "one of the" receptors for these compounds and that ghrelin could be "one" of a number of natural ligands activating the GHS-R (see below). This marks the latest phase of the saga of GHS/ghrelin and its receptors, a receptor family that maintains the same mysterious and magical characteristics of the xenobiotic GHS.

3.   STRATEGIES FOR THE GHS RECEPTOR IDENTIFICATION

3.1   GHS ligands and binding studies

The first report on the binding of met-enkephalin-derived GHRP in rat forebrain appeared in 1988 (5). It revealed that GHRP did not displace [^{3}H]naloxone binding from opiate receptors suggesting that these compounds acted through distinct binding sites. Using [^{125}I]Tyr-Ala-GHRP-6 as a radioligand, Bowers and coworkers identified the presence of specific GHRP binding sites in the rat hypothalamus and pituitary gland showing that different unlabelled peptidyl GHS, but not GHRH, displaced the binding of the radiotracer (6). In 1995, our studies, focusing on the GHRP receptor distribution in the brain, demonstrated for the first time the existence of [^{125}I]Tyr-Ala-hexarelin binding sites in the human hypotalamus-pituitary area, but also remarkable presence of these binding sites in other brain regions, such as choroid plexuses, cerebral cortex, hippocampus and medulla-oblongata suggesting that GHRP are the synthetic counterpart of an endogenous GHS involved in the neuroendocrine control of GH secretion and possibly also in other central activities (40). Afterwards, most pharmacological companies involved in the GH field and several independent groups developed and tested orally active GHRP-6 analogues and with improved potency and bioavailability. One of these developed compounds was MK-0677 (11), a non-peptidyl GHS that was used for binding studies and for cloning the GHS-R. The MK-0677 binding sites in the rat and porcine pituitary were characterized by the use of radiolabelled [^{35}S]MK-0677 (13). Binding was displaced by non-peptidyl and peptidyl GHS, but not by GHRH, SRIF, met-enkephalin, substance P, galanin and other neurotransmitters. Moreover, the relative IC_{50} values for displacement of [^{35}S]MK-0677 binding was highly correlated with the activity of different GHS in stimulating GH release from cultured rat pituitary cells. Binding studies also indicated that the GHS-R was G-protein coupled, because high-affinity binding was inhibited noncompetitively by stable guanosine triphosphate analogues. Following ligand binding, signal transduction is mediated by PLC and IP_3 and this signalling pathway was exploited to clone the receptor by expression cloning techniques (12,14). Recently, the de-orphanization of the GHS-R consequent to the discovery of its natural ligand ghrelin allowed the preparation of some radioiodinated ghrelin derivatives ([^{125}I]Tyr^4-des-Phe^4-ghrelin and [^{125}I]His^9-ghrelin) that have been shown to be reliable probes for labelling GHS-R by *in vitro* binding studies in different animal and human tissues (19, 28-30, 41,42).

3.2   Cloning and transfection techniques

The cloning strategy was based on the findings that the signal transduction used by MK-0667, GHRP-6 and similar compounds appeared to involve PLC resulting in a rise in IP_3 and intracellular Ca^{2+} (43). *Xenopus* oocytes were injected with swine poly(A)+ mRNA, as a source of GHS-R mRNA, supplemented with various G-alpha subunits mRNAs, and Ca^{2+} activated Cl$^-$ currents were assessed after being challenged with MK-0677. Using this approach the Merck investigators identified a single cDNA which encoded the GHS-R (12). After transient transfection of HEK-293 or COS-7 cells with GHS-R cDNA, Ca^{2+}-mediated luminescence signal and [^{35}S]MK-0677 competition binding assays were performed to characterize the pharmacological properties of the cloned receptor (14). Because this receptor binds GHRP in addition to MK-0677 and is activated by these compounds, it was designated the "growth hormone secretagogue receptor" (14).

4.   CLONING THE GHS-RECEPTOR

The GHS-R was initially cloned utilizing a cDNA library pool from swine anterior pituitary (12) where GHS-R is present in low abundance (13). The GHS-R was subsequently identified in the human pituitary and in the hypothalamus of several animal species (14). The GHS-R gene is located on human chromosome 3q26.2 and is composed by two exons encoding a seven transmembrane domain protein and one intron (44). Exon one encodes the amino-terminal extracellular domain, five transmembrane domains, three intracellular loops and two of the three extracellular loops; exon two encodes transmembrane domains 6 and 7, the third extracellular loop and the intracellular carboxyl terminal segment. The receptor is a member of the heptahelical superfamily of G-protein-coupled receptors (GPC-R). Two GHS-R mRNA isoforms, defined as types 1a and 1b are encoded by the GHS-R gene and are produced by alternative mRNA processing. Translation of the GHS-R mRNA isoform, where the intron is removed intact, produces the biological active receptor GHS-R1a, while GHS-R1b mRNA, encodes a C-terminally truncated protein that contains a short intron coding sequence (see below for its biochemical characteristics).

4.1 Molecular characteristics of the GHS-R1a

The human GHS-R1a is a polypeptide of 366 amino acids with a molecular mass of approximately 41 kDaltons (kDa) and is the functional form of the receptor. Based on its deduced peptide sequence, the GHS-R1a is related to the rhodopsin superfamily, but is not related to known subfamilies of GPC-R. The closest relatives are the neurotensin receptor, the thyrotropin-releasing hormone receptor and the motilin receptor type-1a with 59%, 56% and 52% similarity, respectively (14). Comparison of the predicted human rat, pig and sheep GHS-R1a amino acid sequences reveals a high degree of sequence identity between 91.8% and 95.6% and the existence of this receptor can apparently be extended to pre-Cambrian times as amino acid sequences strongly related to human GHS-R1a have been identified in teleost fish, such as the Japanese Pufferfish (*Fugu rubipres*) (45). These observations strongly suggest that the GHS-R1a and its natural ligand ghrelin play a fundamental role in biology. Synthetic GHS and ghrelin, as well as des-Gln[14]-ghrelin, bind with high affinity to the GHS-R1a; their efficacy in displacing [^{35}S]MK-0677 or [^{125}I]Tyr[4]-ghrelin binding from pituitary membranes or the cloned GHS-R1a correlates well with concentrations required to stimulate GH release (13,18,41). By contrast, des-acyl ghrelin, a natural form of ghrelin devoid of GH-releasing activity *in vivo* (19), does not bind GHS-R1a (18) or displace radiolabelled ghrelin from the hypothalamic or pituitary binding sites (19,41). The basic amine common to peptidyl GHRP-6 and non-peptidyl MK-0677 likely establishes an electrostatic interaction with a negatively charged site in the GHS-R1a (46). Ligand amine binding by the GHS-R1a probably takes place at Glu124 in the third transmembrane domain as substitution of glutamine for glutamic acid (Glu124Gln mutant) in human GHS-R1a inactivates receptor function while not altering its three dimensional configuration. Whereas all peptidyl and non-peptidyl GHS share a common binding pocket in the third transmembrane region of GHS-R1a, there are other distinct binding site domains of this receptor that appear to be selective for particular classes of agonists. Met213 is unimportant for the binding of MK-0677, but it is critical to binding of the biphenyl moiety in the more active non-peptide L-692,585 GHS and less critical for the peptide GHRP-6 (47). Finally, the activity of all agonists can be completely abolished mutating Cys116, that form a disulfide bond with Cys198, in Ala (Cys116Ala mutant) in the first extracellular loop of the top of helix 3 (45,46). Actually, the binding site domain of the natural ligand ghrelin, which is structurally unrelated to synthetic GHS, is yet to be experimentally determined. Other likely sites of contact between other endogenous substances and GHS-R1a have been found in other regions of this receptor. For example, adenosine is a partial agonist of the GHS-R1a and binds to a receptor site distinct from the binding pocket recognized by peptidyl and non-peptidyl GHS (48). In contrast, D-Arg^1-D-Phe^5-D-Trp^{7,9}-Leu^{11}-substance P, a substance P antagonist which inhibits the GH release in response to GHS by binding to pituitary ghrelin/GHS-R (7), was found to be a high potency full inverse agonist as it decreased, in GHS-R1a-tranfected COS-7 cells, the constitutive signalling of the ghrelin receptor down to that observed in un-transfected cells (49). The fact that the GHS-R is highly constitutively active raises a series of questions concerning the physiological importance of this activity and whether the receptor is controlled not only by the ghrelin agonist but also by a yet unknown endogenous inverse agonist. Furthermore, we demonstrated (50) that GHS-R is bound also by another endogenous molecule, such as cortistatin (CST), a neuropeptide homologous to SRIF, suggesting that CST, but not SRIF, besides ghrelin and adenosine, could be another natural ligand of the GHS-R possibly involved in modulating the activity of this receptor (see also below). After the binding of the ligand, GHS-R1a acts through the $G_{\alpha11}$ subunit of G-protein to activate PLC resulting in hydrolysis of membrane-bound phosphatidylinositol bisphosphate (PIP$_2$) to generate IP$_3$ and diacylglycerol (DAG). Intracellular free Ca^{2+} concentration increases due to rapid but transient release of Ca^{2+} from IP$_3$-responsive cytoplasmatic storage pools and more sustained inflow of Ca^{2+} due to the activation of L-type calcium channels and blockade of K$^+$ channels resulting in depolarization of the somatotroph and release of GH (14). DAG activates protein kinase C and its messenger systems including adenylate cyclase, thereby secondarily raising intracellular cyclic AMP levels. There are species-specific second messenger pathways activated by various GHS; thus in sheep but not in rat pituitary cells GHRP-2 activates the adenyl cyclase/cyclic AMP/protein kinase A pathway, while GHRP-6 and non peptidyl GHS act through the PLC pathway to enhance GH release (51). These differences might be explained by different conformations of the GHS-R1a induced by each ligand and, as a consequence, the receptor could couple to different G proteins and activate alternative signal transduction pathways or could reflect the possibility of more than one type of GHS-R (see below). It is speculated that ghrelin stimulates GH release via the same pathways, but the exact cellular mechanisms of this ghrelin activity are yet to be elucidated. Recent evidence indicates that ghrelin activates in infant rats the expression of a pituitary-specific transcription· factor (Pit-1) that stimulates GH gene transcription in anterior pituitary cells and therefore could increase GH production in the body via upregulation of Pit-1 expression (52).

4.2   Tissue distribution and functional significance of the GHS-R1a

Expression of the GHS-R1a was shown in the hypothalamus and anterior pituitary gland (14) consistent with its role in stimulating GH release. In human fetuses, where circulating ghrelin levels have been recently documented (53), mRNA for GHS-R1a was already detected at 18- and 31-week gestation, suggesting that ghrelin/GH axis might be active in early development (54). GHS-R1a mRNA is largely confined in somatotroph pituitary cells and in the arcuate nucleus, a hypothalamic area that is crucial for the neuroendocrine and the appetite-stimulating activities of ghrelin and synthetic GHS. This is supported by the demonstration that ghrelin and GHS, effectively stimulate the expression of some markers of neural activity (*c-fos* and early growth response factor-1) in the arcuate nucleus neurons. The activated hypothalamic cells include GHRH-containing neurons, but also cells containing the appetite stimulating neuropeptide Y and an endogenous melanocortin receptor antagonist such as agouti-related protein (20). Less abundant but detectable levels of GHS-R1a mRNA were also demonstrated in other hypothalamic regions (dorsomedial, ventromedial and paraventricular nuclei and in neurons near the median eminence), as well as in various extra-hypothalamic areas, such as the hippocampus, pars compacta of the substantia nigra, ventral tegmental area, dorsal and medial raphe nuclei and pyriform cortex (14,47). The wide expression of the GHS-R1a in the brain is in keeping with observations suggesting for ghrelin and synthetic GHS broader central actions beyond the control of GH secretion and appetite (20,21 and other chapters). Using more sensitive real-time reverse transcriptase polymerase chain-reaction expression, GHS-R1a mRNA was detected in the thyroid gland, stomach, intestine, pancreas, spleen, ventricular myocardium, aorta, lung, adrenal gland, kidney, adipose tissue, testis, ovary, and lymphocytes, whereas no expression was detected, even at a high cycle number, in the atrial myocardium, liver, prostate, bladder, breast, placenta, lymh node, skin, skeletal muscle or vein (20,55,56). Binding studies performed in our laboratory using [^{125}I]Tyr-Ala-hexarelin as a ligand confirmed the presence of GHS-R in the human hypothalamus and pituitary (7). Well detectable binding sites were also found in human choroid plexuses, cerebral cortex, hippocampus and medulla-oblongata, whereas negligible binding ws observed in the thalamus, striatum, substantia nigra, cerebellum and corpus callosum (7). In another binding study (35) of a wide range of human tissues using the same ligand, GHS-R was detected mainly in the myocardium, but they were also present (in order of decreasing binding activity), with no sex-related differences, in adrenal gland, testis, arteries, lung, ovary, liver, skeletal muscle, kidney, thyroid gland, adipose tissue, veins, uterus, skin and lymph nodes. In contrast, negligible binding was found in parathyroid, salivary gland, stomach, colon, pancreas, spleen, breast, placenta and prostate. The presence of GHS-R1a mRNA and GHS binding sites in a large number of peripheral tissues explains the wide range of GH-independent activities of ghrelin (i.e.: gastric acid-stimulating and gastrointestinal prokinetic actions; anti-lipolitic effects; modulation of the immune system and regulation of adrenal and gonadal steroidogenesis; influence on pancreatic functions and glucose metabolism; vasodilator and cardiotropic activities) and indicates that ghrelin is much more that a natural GH secretagogue and then a novel pleiotropic hormone (20, 21,23,55-59 and other chapters). The discrepancy between mRNA analysis and binding studies points to the existence of additional GHS binding sites (see below). Studies on human pituitary tumors revealed significant GHS-R1a expression in somatotroph, lactotroph, and corticotroph adenomas, but absent or low levels in most thyrotroph, gonadotroph and silent adenomas. Expression of GHS-R1a was also found in human prostate cancer cell lines, in human endocrine bronchial tumors and in other neurondocrine neoplasms of stomach, pancreas and intestine (20,60 and Chapter 10). Similarly, the presence of GHS-R was also demonstrated in human breast and thyroid carcinomas, as well as in non-endocrine neoplasms of lung tissues (20,27,36). Therefore, the ghrelin/GHS-R system may have a role in the control of neoplastic cell survival and proliferation (see Chapter 10). Recently, [^{125}I]His^{9}-ghrelin binding sites have been detected in human surgical samples, localizing to the smooth muscle layer of aorta, coronary artery and saphenous vein, with three-to-four up-regulation of receptor density in atherosclerotic coronary arteries and saphenous vein grafts with advanced intimal thickening suggesting that ghrelin may have a role in such pathological processes (42).

5.   MULTIPLE NON-TYPE 1A GHS-R SUBTYPES

There is a growing body of evidence suggesting that the cloned GHS-R is not "the" sole receptor, but "one of the" receptors for this family of compounds. This concept is supported by cloning the GHS-R1b isoform and various GHS-R1a homologs, by the evident differences in the binding activities of natural (ghrelin and its des-acylated form), synthetic peptidyl (hexarelin and its analogues), and non-peptidyl (MK-0677) molecules, by the demonstration that not all GHS show the same biological activities and by the fact that knockout mice and female rats genetically deficient of GHS-R1a exhibit a normal growth rate. Furthermore, there is ongoing controversy as to whether ghrelin is "the" sole ligand or "one of a number" of endogenous ligands, because other endogenous substances have been shown to interact with the GHS-R (see below).

5.1   The spliced GHS-R1b form and the GHS-R1a homologs

The GHS-R1b is a splice variant of the GHS-R1a. GHS-R1b is truncated and consists of a 298 amino acid protein containing only the first five transmembrane domains plus a unique 24 amino acid "tail" encoded by alternative spliced intronic sequence. In GHS-R1b transfected cells treated with the synthetic GHS, the GHS-R1b failed to bind GHS and to respond to these compounds and thus, this receptor isoform is not known to exibit any biological activity (14). However, no binding or functional studies have been reported with the endogenous GHS-R1a ligand ghrelin and given the wide spread expression of GHS-R1b in many normal and neoplastic GHS-R1a positive or negative tissues (61 and Chapter 10), GHS-R1b has yet to be shown to have functional significance, yet three GHS-R1a homologs (GPR38, GPR39 and FM3) have been isolated from human genomic libraries. The deduced amino acid sequence of one of these clones, GPR38, that has been recently identified as the motilin receptor (62), has 52% identity and 73% similarity to the human GHS-R1a (48). Although motilin was reported to have structural similarities with ghrelin and to share with it some biological activities (GH-releasing, intestinal prokinetic and orexigenic effects), it does not activate the GHS-R1a (48,63). Thus, ghrelin and motilin represent a novel family of gastrointestinal peptides with overlapping gut-brain functions that act through homolog receptors requiring the continued presence of their respective ligands to be activated (63).

5.2 GHS binding sites common for ghrelin and des-acyl ghrelin

Our group, working in collaboration with A. Graziani in Novara, demonstrated for the first time that ghrelin, as well as peptidyl and non-peptidyl GHS and even GHS analogues devoid of any GH-releasing activity, prevent cell death of cultured cardiomyocytes and endothelial cells induced by either doxorubicin, serum withdrawal or activation of FAS, stimulating survival intracellular signalling pathways such as tyrosine phosphorilation of intracellular proteins and activation of extracellular-signal-regulated kinase-1 and -2 and kinase B/AKT (28). Interestingly, the same effects of acylated ghrelin are shared by the des-acyl molecule, thus indicating that the acylation of the peptide is needed for the endocrine activities and that even the des-acylated ghrelin is a biologically active peptide. Since des-acyl ghrelin is generally unable to bind the GHS-R1a and the pituitary GHS binding sites (18) (Figure 3-1A), this evidence would imply the existence of another GHS-R subtype, common for ghrelin and its des-acylated form, that could mediate an anti-apoptotic effect in the cardiovascular system (28). In agreement with these observations, more recent data in our laboratory (29) indicate that ghrelin either in acylated or des-acylated form, as well as peptidyl GHS, show similar negative inotropic effect on isolated guinea pig papillary muscle. The potency of these compounds (des-acyl ghrelin>ghrelin>hexarelin) is consistent with binding experiments performed on ventricular membranes, where either des-acylated or acylated ghrelin forms and hexarelin compete with $[^{125}I]Tyr^4$-ghrelin for a common binding site (Figure 3-1C). A similar pattern of displacement was observed by us in the human pancreas (Figure 3-1D), where des-acyl ghrelin has been shown to reverse the effects that ghrelin exerts on the insulin secretion in humans (personal unpublished results).

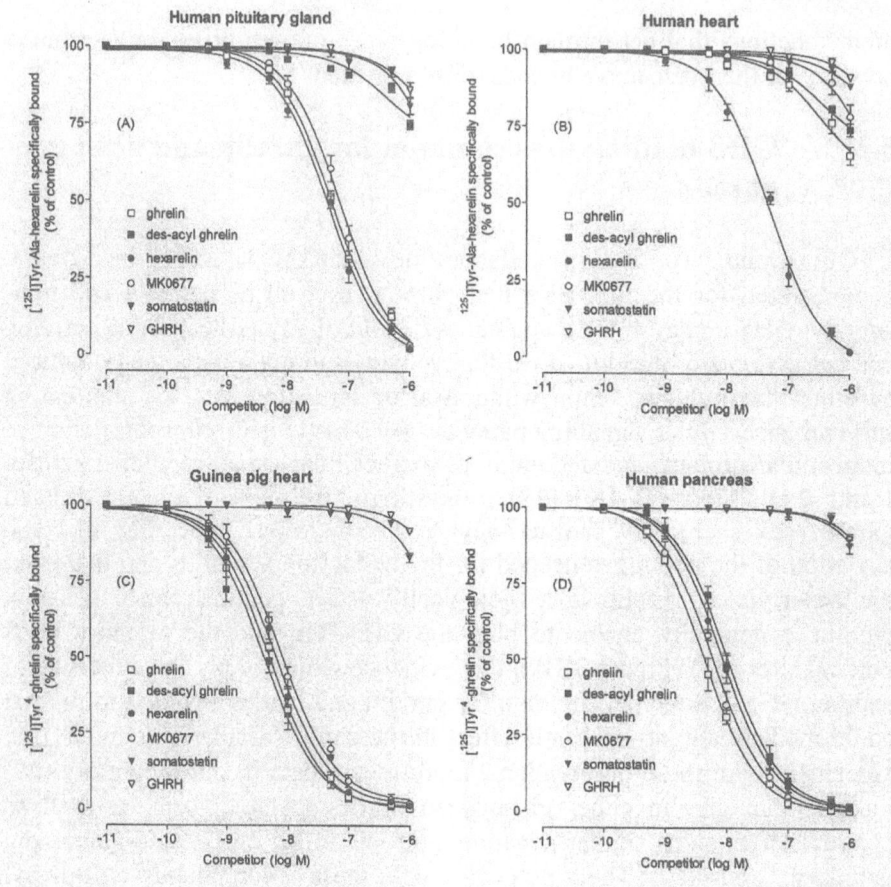

Figure 3-1. Displacement of [^{125}I]Tyr-Ala-hexarelin from membranes of pituitary (A) and human heart (B) and of [^{125}I]Tyr4-ghrelin from membranes of guinea pig heart (C) and human pancreas (D) by different unlabeled competitors. The ordinate represents binding as a percentage of control (specific binding in the absence of unlabeled competitors). Values are means ± standard error of the mean of three separate experiments.

Interestingly, GHS-R has been found even in tumoral tissues from organs which do not express these receptors in physiological conditions such as the breast. The presence of specific GHS-R was shown in breast cancer, but not in fibroadenomas and normal mammary parenchyma (27). In breast tumors, the highest binding activity is present in well-differentiated invasive breast carcinomas and is progressively reduced in moderately- to poorly-differentiated tumors. GHS-R are also present in both estrogen dependent (MCF7 and T47D) and independent (MDA-MB231) breast cancer cell lines, in which ghrelin, synthetic GHS and EP-80317, a hexarelin analog devoid of any GH-releasing properties *in vivo*, cause inhibition of cell proliferation at concentrations close to their binding affinity. Like in the cardiovascular system, the same effect of acylated ghrelin is shared by the des-acylated molecule further indicating that des-acyl ghrelin is a biological active peptide possessing also antiproliferative actions. This evidence also supports the possibility that a specific binding site, common for ghrelin and des-acyl ghrelin, seems to mediate the antiproliferative effects of these substances. It is not yet known if this receptor is the same of that found in the heart or represents another additional subtype. These data indicate that a GHS-R shared by ghrelin, des-acyl ghrelin and synthetic GHS is also present in human prostate neoplasms and related cell lines (30). Like in breast tumors, this prostatic GHS-R seems to be involved in the control of neoplastic cell survival and proliferation.

5.3 GHS binding sites not shared by ghrelin

Ghrelin and all the GHS developed so far seem to exhibit a similar high affinity to GHS-R1a. However, there is strong evidence suggesting the existence of additional GHS-R subtypes which may exhibit different affinities for these compounds. In particular, specific binding sites for Tyr-Ala-hexarelin and other peptidyl GHS (GHRP-2, GHRP-6 and hexarelin), with a density more remarkable than that of GHS-R1a, have been found in some endocrine glands (pituitary, thyroid and adrenal), as well as in a wide range of non-endocrine peripheral animal and human tissues such as heart, lung, arteries, skeletal muscle, kidney, liver, uterus and adipose tissue (31-36). These binding sites are presumably different from the GHS-R1a or GHS binding sites found in endocrine glands, because they show a very low binding affinity for ghrelin and the non-peptidyl GHS MK-0677. This is illustrated in Figure 3-1, which compares the ability of different unlabeled competitors to displace [^{125}I]Tyr-Ala-hexarelin binding to membranes from human pituitary gland (Figure 3-1A) and heart (Figure 3-1B). Studies utilizing a photoactivable hexarelin (p-benzoyl-Tyr-Phe-hexarelin) suggested a second distinct GHS-R subtype in human and bovine pituitary with a molecular weight of 57 kDa (33) and a third subtype in rat heart with a molecular weight of 84 kDa (37). Data obtained by us and by Locatelli's group in Milan with the tripeptide EP-51389 developed by R. Deghenghi in Europeptides are consistent with this view. EP-51389 is as effective as hexarelin in stimulating GH secretion in the rat, but it is far less effective in protecting the heart from ischemia (39). Interestingly, EP-51389 effectively displaced hexarelin from its hypothalamic binding sites, but poorly from cardiac membranes (64). More recently, Ong and coworkers (65) identified this cardiac binding site as CD36, a multifunctional class B scavenger receptor that is expressed in many tissues, among them microvascular endothelium, skeletal and smooth muscle cells, and monocytes/macrophages. It has been implicated in multiple physiological functions (i.e. cellular adhesion, fatty acid and lipid transportation and antigen presentation) and pathological processes related to macrophage foam cell formation and the pathogenesis of atherosclerosis. The recent report by Ong research group also demonstrates an unexpected vasocostrictive activity elicited by hexarelin in perfused heart preparation. Hexarelin increases the coronary perfusion pressure by direct interaction with CD36 but is devoid of any effect in CD36 knockout mice and rats genetically deficient of CD36. This cardiovascular effect of hexarelin appears distinct from that of ghrelin and MK-0677 that do not share all the cardiotropic actions of peptidyl GHS (39,66,67). Thus, ghrelin, hexarelin and MK-0677 would have different cardiovascular effects based mainly on their ability to bind multiple GHS-R subtypes common for hexarelin, MK-0677 and various ghrelin forms (28,29, see also above), specific for ghrelin and hexarelin only (66,67) or specific (CD36) for hexarelin alone (32,35,37). Each receptor subtype could then contribute independently to the wide array of cardiovascular activities induced by synthetic GHS and ghrelin. Recently, we demonstrated that GHS-R, exhibiting a binding profile different from the GHS-R1a, are present in human lung tumors and related cancer cell lines, in which synthetic peptidyl GHS, but not ghrelin, cause inhibition of cell growth in vitro (36). It is likely that further subtypes of GHS-R will be cloned, since there is evidence pointing toward the possible existence of additional ghrelin /GHS-R subtypes different from the already cloned one and the CD36 receptor.

6.   POSSIBLE EXISTENCE OF OTHER NATURAL GHS-R LIGANDS

Recently, particular interest has been given to the identification of new natural ligands of the GHS-R endowed with agonist or antagonist activity (see Table 3-1). Investigators of Novo Nordisk group working on HEK-293 cells expressing human or pig GHS-R1a found that also adenosine activates the transfected receptor but does not possess a biological counterpart being unable to stimulate GH secretion and amplify the GHRH effects on normal pituitary cell cultures (68). It has been demonstrated that adenosine is a partial agonist of the pituitary GHS-R1a and binds to a receptor site distinct from the binding pocket recognized by GHRP-6 and MK-0677 (48).

Table 1. Synthetic and natural ligands of GHS-R

Synthetic ligands	Natural ligands	Receptor activity	References
Peptidyl and non-peptidyl GHS	ghrelin and des-Gln[14]-ghrelin	agonist[a]	15,18,20,24
D-Lys[3]-GHRP6	///	antagonist[a]	7,15
EP-80317[b]	des-acyl ghrelin	agonist[f]	27-30
R-PIA[c]	adenosine	partial agonist[g]	48,68
EP-01492[d]	cortistatin-14 and -17	unknown	50,70
Substance P antagonist H-6935[e]	unknown	inverse agonist[h]	49

[a]of the GHS-R1a; [b]Haic-D-2Me-Trp-Lys-Trp-D-Phe-Lys-NH$_2$; [c]A$_1$ receptor agonist N^6-R-phenylisopropyladenosine; [d]cortistatin-8: Pro-c[Cys-Phe-D-Trp-Lys-Thr-Cys]-Lys; [e]D-Arg^1-D-Phe^5-D-Trp^{7,9}-Leu^{11}-substance P; [f]on non-type 1a GHS-R of breast cancer cell lines and cardiomyocytes; [g]on intracellular Ca^{2+} concentration in HEK-293 cells expressing GHS-R1a; [h]on phospholipase C activity of COS-7 cells expressing GHS-R1a.

These observations are intriguing and perhaps could explain, at least in part, the reported effects of hexarelin on the heart, where adenosine plays an important role in regulating cardiac functions (69). However, the reciprocal interferences between adenosine, GHS and their cardiac receptors have not been examined to date and remain still unexplored. As described in the previous sections, even des-Gln^{14}-ghrelin and des-acyl ghrelin bind the GHS-R and are biological active. Therefore, these natural ghrelin analogues can be considered as additional endogenous ligands of the GHS-R. Furthermore, we reported (50) that GHS-R is bound also by another endogenous molecule such as CST, a neuropeptide homologous to SRIF which in turn is unable to recognize GHS-R. Our finding supports the hypothesis that other natural ligands, such as CST could modulate the activity of GHS-R and that this neuropeptide is not simply a natural SRIF analogue (70). Finally, the recent demonstration (49) that D-Arg^1-D-Phe^5-D-Trp^{7,9}-Leu^{11}-substance P, a synthetic high-affinity ligand of GHS-R (7), induces signalling effects opposite to ghrelin (49), suggests the possible existence of a yet unknown endogenous inverse agonist that could modulate or fine-tune a high level of constitutive ghrelin activity.

ACKNOWLEDGEMENTS

We gratefully acknowledge the participation to the original studies reported of Drs. Giuseppe Alloatti, Emanuela Arvat, Paola Cassoni, Corrado Ghè, Andrea Graziani, Jorgen Isgaard, Vittorio Locatelli, Huy Ong, Mauro Papotti and Riccardo Zucchi. We also thank Prof. Romano Deghenghi for its invaluable and enthusiastic contributions for both basic and clinical aspects of this study. The present review was supported by grants from the Italian Ministry of Education and University (MIUR, Rome) (ex-60% and Cofin 2002 to G. Muccioli and E. Ghigo) and the Fondazione per lo Studio delle Malattie Endocrine e Metaboliche (SMEM, Turin).

REFERENCES

1. Bowers CY, Momany F, Chang D, Hong A. Chang K. Structure-activity relationships of a synthetic pentapeptide that specifically releases GH in vitro. Endocrinology. 1980; 106:663-70.
2. Bowers CY, Momany F, Reynolds GA, Hong A. On the in vitro and *in vivo* activity of a new synthetic hexapeptide that acts on the pituitary to specifically release growth hormone. Endocrinology. 1984; 114:1537-45.
3. Bercu BB, Yang SW, Masuda R, Walker RF. Role of selected endogenous peptide in growth hormone-releasing hexapeptide activity: analysis of growth hormone-releasing hormone, thyroid hormone-releasing hormone, and gonadotropin-releasing hormone. Endocrinology. 1992; 130:2579-86.
4. Camanni F, Ghigo E, Arvat E. Growth hormone-releasing peptides and their analogs. Front Neuroendocrinol. 1998; 19:47-72.
5. Codd EE, Yellin T, Walker RF. Binding of growth hormone-releasing hormone and enkephalin-derived growth hormone releasing peptides to mu and delta opioid receptors in forebrain of rat. Neuropharmacology. 1988; 27:1019-25.
6. Sethumadavan K, Veeraragavan K, Bowers CY. Demonstration and characterization of the specific binding of growth hormone-releasing peptide to rat anterior pituitary and hypothalamus. Biochem Biophys Res Commun. 1991; 178:31-7.
7. Muccioli G, Ghè C, Ghigo MC, et al. Specific receptors for synthetic GH secretagogues in the human brain and pituitary gland. J Endocrinol. 1998; 157:99-106.
8. Bowers CY, Reynolds GA, Durham D, Barrera CM, Pezzoli SS, Thorner MO. Growth hormone(GH)-releasing peptide stimulates GH release in normal men and acts synergistically with GH-releasing hormone. J Clin Endocrinol Metab. 1990; 70:975-82.
9. Deghenghi R, Cananzi M, Torsello A, Battisti C, Muller EE. GH-releasing activity of hexarelin, a new growth hormone-releasing peptide, in infant and adult rats. Life Sci. 1994; 54:1321-8.
10. Guerlavais V, Boeglin D, Mousseaux D, et al. New active series of growth hormone secretagogues. J Med Chem. 2003; 46:1191-203.
11. Smith RG, Cheng K, Pong S-S, et al. A nonpeptidyl growth hormone secretagogue. Science. 1993; 260:1640-3.
12. Howard AD, Feighner SD, Cully DF, et al. A receptor in pituitary and hypothalamus that functions in growth hormone release. Science. 1996; 273:974-7.

13. Pong SS, Chaung LY, Dean DC, Nargund RP, Patchett AA, Smith RG. Identification of a new G-protein-linked receptor for growth hormone secretagogues. Mol Endocrinol. 1996; 10:57-61.

14. Smith RG, van der Ploeg LN, Howard AD, et al. Peptidomimetic regulation of growth hormone secretion. Endocr Rev. 1997; 18:621-45.

15. Kojima M, Hosoda H, Date Y, Nakazato M, Matsuo H, Kangawa K. Ghrelin is a growth hormone-releasing acylated peptide from stomach. Nature. 1999; 402:656-60.

16. Arvat E, Maccario M, Di Vito L, et al. Endocrine activities of ghrelin, a natural growth hormone secretagogue (GHS), in humans: comparison and interactions with hexarelin, a nonnatural peptidyl GHS. J Clin Endocrinol Metab. 2001; 86:1169-74.

17. Banks WA, Tschop M, Robinson SM, Heiman ML. Extent and direction of ghrelin transport across the blood-brain barrier is determined by its unique primary structure. J Pharmacol Exp Ther. 2002; 302:822-7.

18. Bednarek MA, Feighner SD, Pong S-S, et al. Structure-functions studies on the new growth hormone-releasing peptide, ghrelin: minimal sequence of ghrelin necessary for activation of growth hormone secretagogue receptor 1a. J Med Chem. 2000; 43:4370-6.

19. Torsello A, Ghè C, Bresciani E, et al. Short ghrelin peptides neither displace ghrelin binding *in vitro* nor stimulate GH release *in vivo*. Endocrinology. 2002; 143:1968-71.

20. Muccioli G, Tschop M, Papotti M, Deghenghi R, Heiman M, Ghigo E. Neuroendocrine and peripheral activities of ghrelin: implications in metabolism and obesity. Eur J Pharmacol. 2002; 440:235-54.

21. Broglio F, Gottero C, Arvat E, Ghigo E. Endocrine and non-endocrine actions of ghrelin. Horm Res. 2003; 59:109-17.

22. Horvath TL, Diano S, Tschop M. Ghrelin in hypothalamic regulation of energy balance. Curr Top Med Chem. 2003; 3:921-7.

23. Lazarczyk MA, Lazarczyk M, Grzela T. Ghrelin: a recently discovered gut-brain peptide (Review). Int J Mol Med. 2003; 12:279-87.

24. Hosoda H, Kojima M, Matsuo H, Kangawa K. Purification and characterization of rat des-Gln[14]-ghrelin, a second endogenous ligand for the growth hormone secretagogue receptor. J Biol Chem. 2000; 275:21995-2000.

25. Hosoda H, Kojima M, Mizushima Y, Shimizu S, Kangawa K. Structural divergence of human ghrelin. Identification of multiple ghrelin-derived molecules produced by post-translational processing. J Biol Chem. 2003; 278:64-70.

26. Hosoda H, Kojima , Matsuo H, Kangawa K. Ghrelin and des-octanoyl ghrelin: two major forms of rat ghrelin peptide in gastrointestinal tissue. Biochem Biophys Res Commun. 2000; 279;909-13.

27. Cassoni P, Papotti M, Ghè C, et al. Identification, characterization and biological activity of specific receptors for natural (ghrelin) and synthetic growth hormone secretagogues in human breast carcinomas and cell lines. J Clin Endocrinol Metab. 2001; 86:1738-45.

28. Baldanzi G, Filigheddu N, Cutrupi S, Ghrelin and des-acyl ghrelin inhibit cell death in cardiomyocytes and endothelial cells through ERK1/2 and PI 3-kinase/AKT. J Cell Biol. 2002; 159:1029-37.

29. Bedendi I, Alloatti G, Marcantoni A, et al. Cardiac effects of ghrelin and its endogenous derivatives des-octanoyl ghrelin and des-Gln[14]-ghrelin. Eur J Pharmacol. 2003; 476:87-95.

30. Cassoni P, Ghè C, Allia E, et al. Expresion of ghrelin and biological activity of specific receptors for ghrelin and des-acyl ghrelin in human prostate neoplasms and related cell lines. (submitted for publication).

31. Muccioli G, Papotti M, Catapano F, et al. Specific receptors for synthetic growth hormone secretagogues in human tissues. Arch Pharm. (Weinheim) 1998; 358 (Suppl 2):R548.

32. Muccioli G, Papotti M, Ong H, et al. Presence of specific receptors for synthetic growth hormone secretagogues in the human heart. Arch Pharm. (Weinheim) 1998; 358 (Suppl 2):R549.

33. Ong H, McNicoll N, Escher E, et al. Identification of a pituitary growth hormone-releasing peptide (GHRP) receptor subtype by photoaffinity labeling. Endocrinology. 1998; 139:432-5.

34. Ong H, Bodart V, McNicoll N, Lamontagne D, Bouchard JF. Binding sites for growth hormone-releasing peptide. Growth Hormone IGF Res. 1998; 8:137-40.

35. Papotti M, Ghè C, Cassoni P, et al. Growth hormone secretagogue binding sites in peripheral human tissues. J Clin Endocrinol Metab. 2000; 85:3803-7.

36. Ghè C, Cassoni P, Catapano F, et al. The antiproliferative effect of synthetic peptidyl GH secretagogues in human CALU-1 lung carcinoma cells. Endocrinology. 2002; 143:484-91.

37. Bodart V, Bouchard JF, McNicoll N, et al. Identification and characterization of a new growth hormone-releasing peptide receptor in the heart. Circ Res. 85:796-808.

38. Arvat E, Broglio F, Aimaretti G, et al. Ghrelin and synthetic GH secretagogues. Best Pract Res Clin Endocrinol. 2002; 16:505-17.

39. Torsello A, Bresciani E, Rossoni G, et al. Ghrelin plays a minor role in the physiological control of cardiac function in the rat. Endocrinology. 2003; 136:1146-52.

40. Muccioli G, Ghè C, Ghigo MC, et al. Presence of specific receptors for hexarelin, a GH-releasing hexapeptide in human brain and pituitary gland. 20th International Symposium on Growth Hormone and Growth factors in Endocrinology and Metabolism, Berlin, 1995; p 110 (abstract).

41. Muccioli G, Papotti M, Locatelli V, Ghigo E. Binding of ^{125}I-labeled ghrelin to membranes from human hypothalamus and pituitary gland. J Endocrinol Invest. 2001; 24:RC7-RC9.

42. Katugampola SD, Kuc RE, Maguire JJ, Davenport AP. G-protein-coupled receptors in human atherosclerosis: comparison of vasoconstrictors (endothelin and thromboxane) with recently de-orhanized (urotensin II, apelin and ghrelin) receptors. Clin Sci. (Lond) 2002; 103(Suppl 48):171S-5S

43. Herington J, Hille B. Growth hormone-releasing hexapeptide elevates intracellular calcium in rat somatotrophs by two mechanisms. Endocrinology. 1994; 135:1100-8.

44. McKee KK, Palyha OC, Feighner SD, et al. Molecular analysis of rat pituitary and hypothalamic growth hormone secretagogue receptors. Mol Endocrinol. 1997; 11:415-23.

45. Palyha OC, Feighner SD, Tan CP, et al. Ligand activation domain of human orphan growth hormone (GH) secretagogue receptor (GHS-R) conserved from pufferfish to humans. Mol Endocrinol. 2000; 14: 160-9.

46. Feighner SD, Howard AD, Prendergast K, et al. Structural requirements for the activation of the human growth hormone secretagogue receptor by peptide and non-peptide secretagogues. Mol Endocrinol. 1998; 12:137-45.

47. Smith RG, Palyha OC, Feighner SD, et al. Growth hormone releasing substances: types and their receptors. Horm Res. 1999; 51(Suppl 3):1-8.

48. Smith RG, Leonard R, Bailey ART, et al. Growth hormone secretagogue receptor family members and ligands. Endocrine. 2001; 14:9-14.

49. Holst B, Cygankiewicz A, Halkjar JT, Ankersen M, Schwartz TW. High constitutive signaling of the ghrelin receptor-identification of a potent inverse agonist. Mol Endocrinol. 2003; 17:2201-10.

50. Deghenghi R, Papotti M, Ghigo E, Muccioli G. Cortistatin, but not somatostatin, binds to growth hormone secretagogue (GHS) receptors of human pituitary gland. J Endocrinol Invest. 2001; 24:RC1-RC3.

51. Chen C. Growth hormone secretagogue actions on the pituitary gland: multiple receptors for multiple ligands. Clin Exp Pharmacol Physiol. 2000; 27:323-9.

52. Garcia A, Alvarez CV, Smith RG, Dieguez C. Regulation of Pit-1 expression by ghrelin and GHRP-6 through the GH secretagogue receptor. Mol Endocrinol. 2001; 15:1484-95.

53. Cortellazzi D, Cappiello V, Morpurgo PS, et al. Circulating levels of ghrelin in human fetuses. Eur J Endocrinol. 2003; 149:111-6.

54. Shimon I, Yan X, Melmed S. Human fetal pituitary expresses functional growth hormone-releasing peptide receptors. J Clin Endocrinol Metab. 1998; 83:174-8.

55. Wang G, Lee H-M, Englander E, Greeley GH Jr. Ghrelin-not just another stomach hormone. Regul Pept. 2002; 105:75-81.

56. Petersenn S, Structure and regulation of the growth hormone secretagogue receptor. Minerva Endocrinol. 2002; 27:243-56.

57. Casanueva FF, Dieguez C. Ghrelin: the link connecting growth with metabolism and energy homeostasis. Rev Endocr Metab Disord. 2002; 3:325-38.

58. Broglio F, Arvat E, Benso A, et al. Ghrelin, much more than a natural growth hormone secretagogue. Isr Med Assoc J. 2002; 4:607-13.

59. Petersenn S, Growth hormone secretagogues and ghrelin: an update on physiology and clinical relevance. Horm Res. 2002; 58 (Suppl 3):56-61.

60. Jeffery PL, Herington AC, Chopin LK. The potential autocrine/paracrine roles of ghrelin and its receptor in hormone-dependent cancer. Cytokine Growth Factor Rev. 2003; 14:113-22.

61. Gnanapavan S, Kola B, Bustin SA, et al. The tissue distribution of the mRNA of ghrelin and subtypes of its receptor. J Clin Endocrinol Metab. 2002; 87:2988-91.

62. Feighner SD, Tan CP, McKee KK, et al. Receptor for motilin identified in the human gastrointestinal system. Science 1999; 284:2184-8.

63. Folwaczny C, Chang JK, Tschop M. Ghrelin and motilin: two sides of one coin? Eur J Endocrinol. 2001; 144:R1-R3.

64. Deghenghi R. Structural requirements of growth hormone secretagogues. In: Bercu BB, Walker RF, eds. Growth hormone secretagogues in clinical practice. New York: Marcel Dekker Inc. 1998; 27-33.

65. Bodart V, Febbraio M, Demers A, et al. CD36 mediates the cardiovascular action of growth hormone-releasing peptides in the heart. Circ Res. 2002; 90:844-9.

66. Pettersson I, Muccioli G, Granata R, et al. Natural (ghrelin) and synthetic (hexarelin) GH secretagogues stimulate H9c2 cardiomyocyte cell proliferation. J Endocrinol. 2002; 175:201-9.

67. Frascarelli S, Ghelardoni S, Ronca-Testoni S, Zucchi R. Effect of ghrelin and synthetic growth hormone secretagogues in normal and ischemic rat heart. Basic Res Cardiol. 2003; 89:401-5.

68. Tullin S, Hansen BS, Ankersen M, Moller J, Von Cappelen KA, Thim L. Adenosine is an agonist of the growth hormone secretagogue receptor. Endocrinology. 2000; 141:3397-402.

69. Obata T. Adenosine production and its interaction with protection of ischemic and reperfusion injury of the myocardium. Life Sci. 2002; 71:2083-103.

70. Deghenghi R, Broglio F, Prodam F, et al. Cortistatin, not simply a natural somatostatin analogue. In: Muller EE, ed. Peptides and non peptides of neuroendocrine and oncologic relevance. From basic to clinical research. Heidelberg: Springer-Verlag. 2003;57-64.

Chapter 4

GHRELIN: A NEW LINK BETWEEN THE NEUROENDOCRINE CONTROL OF THE GH AXIS, FOOD INTAKE AND SLEEP

Virginie Tolle & the INSERM-ATC Nutrition Research Group on Ghrelin[*]
Institut National de la Santé et de la Recherche Médicale INSERM U549, 2 ter rue d'Alésia, 75014 Paris, France

Abstract: Since its discovery in 1999 as an endogenous growth hormone (GH) secretagogue, diverse functions have been attributed to ghrelin, from gastrointestinal functions to cardiac physiology. Nowadays, the main interest has shifted from the GH releasing to the orexigenic properties of the peptide and its ability to increase adiposity. The purpose of the present chapter is to review the mechanisms by which ghrelin may act to regulate GH secretion, food intake and sleep-wake rhythmicity. The complex changes in circulating ghrelin levels in clinical conditions of GH hypo- or hypersecretion in relation with variable energy balance conditions are also reviewed and discussed.

Key words: ghrelin, growth hormone, feeding behavior, sleep

1.   INTRODUCTION

The discovery of ghrelin is certainly one of the best examples of what is now known as Reverse Pharmacology. Indeed, ghrelin was purified from rat stomach as a 28 amino acid peptide with a very unusual n-octanoylation on a serine in position 3 (1). Ghrelin was identified as an endogenous ligand of the formerly orphan growth hormone (GH) secretagogue (GHS) receptor

[*] INSERM-ATC Nutrition Research Group on Ghrelin: Marie Thèrése Bluet-Pajot, Marie Hélène Bassant, Philippe Zizzari, Frédérique Poindessous-Jazat, Catherine Tomasetto, Marie Christine Rio, Bruno Estour, Christine Foulon, Julien Daniel Guelfi, Roland Dardennes, Yves Le Bouc, Gloria S. Tannenbaum, Jacques Epelbaum.

(GHS-R) type 1 (2). The receptor itself had been isolated from a porcine pituitary cDNA library, three years before ghrelin characterization, by expression screening using the synthetic GHS L-163,191 (MK-0677) (3). MK-0677 was one of the last members of this family of peptides and peptidomimetics, born twenty years before, when the pentapeptide DTrp2-Met enkephalin-NH2 was firstly shown to stimulate GH release from pituitary *in vitro* (4). Thus, besides the two antagonistic hypothalamic neurohormones, GH-releasing hormone (GHRH) and somatostatin (SRIF), ghrelin may be a third endogenous factor involved in the regulation of ultradian pulsatile GH secretion. In addition to their GH-releasing properties, two other actions of ghrelin and pharmacologically-designed GHS involve a central action: a) they rapidly and potently stimulate food intake and weight gain mainly through an increase in adipose tissue ((5) and Chapter 7); b) they may participate in sleep regulation.

2.   GHRELIN AND GHS-R EXPRESSION

If there is no doubt that GHS-R are expressed in the hypothalamus-pituitary axis, the situation is much less clear in the case of ghrelin itself.

2.1   Ghrelin expression

Ghrelin is predominantly synthesized in the enteroendocrine cells of the gastrointestinal tract (1,6), ghrelin-positive cells being more abundant in the oxyntic mucosa of the stomach. When assayed with a C-terminal antibody which recognizes total ghrelin immunoreactivity, its concentrations in hypothalamus and pituitary are, respectively, a thousand- and two hundred-fold lower than in the stomach. When assayed with a N-terminal antibody which recognizes only bioactive forms, they are undetectable in hypothalamus and pituitary but present in the stomach (7). Nevertheless, some authors, using N-terminal antibodies, report on the immunocytochemical localization of ghrelin in a few arcuate nucleus (ARC) neurons in the rat brain (8) and, using C-terminal antibodies, in the human brain (9). In this latter study, ghrelin immunoreactivity was also measured in pMolar quantities, by specific radio-immuno assays, in human pituitary and pituitary tumors, but not in hypothalamic samples (9). At any rate, hypothalamic ghrelin production does not substantially contribute to circulating ghrelin concentrations in humans, as evidenced from petrosal sinus measurements and from the lack of central over peripheral ghrelin gradient in two acromegalic subjects undergoing surgery (10).

2.2   GHS-R expression

In the rat, GHS-R mRNA is prominently expressed in ARC and ventromedial nucleus (VMN) but not in the periventricular nucleus (PeV) of the hypothalamus (11,12). Little or no specific hybridization is observed in the pituitary under the conditions that give strong signals in the hypothalamus. No sex difference in GHS-R expression is apparent in ARC while expression in VMN is lower in males than in females (11). ARC GHS-R expression is modulated by GH, being markedly increased in GH-deficient dw/dw dwarf rats, and decreased in dw/dw rats treated with bovine GH. Similar changes are observed in GHRH expression, whereas neuropeptide Y (NPY) expression is reduced in dw/dw rats and increased by bovine GH treatment (11).

Initial quantification of chromogenic and autoradiographic *in situ* hybridization double-labeled cells in male Sprague-Dawley rats revealed that 27% and 22% of GHRH mRNA-containing neurons in ARC and VMN, respectively, expressed the GHS-R (12). A previous study restricted to ARC reported a similar percentage (20-25%) in adult female Wistar rats (13). Thus, ghrelin may directly modulate GHRH release into hypophyseal portal blood, and thereby influence GH secretion through interaction with the GHS-R on GHRH-containing neurons. In male Sprague-Dawley rats, a weak hybridization signal was detected in a minority of SRIF mRNA-containing neurons in PeV, the primary source of hypophysiotrophic SRIF neurons projecting to the median eminence, suggesting that SRIF cells are only minor direct targets for ghrelin's actions on GH (14). The largest proportion (30%) of ARC GHS-R expressing cells is composed of GABAergic neurons expressing NPY and possibly agouti-related peptide (AgRP), two potent orexigenic peptides. However, the percentage of NPY/AgRP and SRIF neurons coexpressing GHS-R mRNA (94% and 30%, respectively) seems far greater in female Wistar than in male Sprague-Dawley rats. These quantitative differences might be due to strain or gender differences or quantification methods. At any rate, in both genders, NPY ARC neurons are those which express GHS-R in the greatest proportion, either in terms of peptide expressing neurons or of GHS-R expressing neurons.

By *in situ* hybridization, the other rat brain areas that display localized and discrete signals for the GHS-R include the CA2, CA3 and dentate gyrus regions of the hippocampus, the medial amygdaloid nucleus, the substantia nigra, the ventral tegmental area, and the dorsal and median raphe nuclei (12,15). In resemblance to the results from rat brain, specific signals in human pituitary, hypothalamus and hippocampus are observed by RNase protection assays. However, the comparative distribution of GHS-R mRNA in the hypothalamus and pituitary of a small primate *microcebus murinus* and in the rat revealed interesting differences between these two species (16). In both species GHS-R expression is particularly dense in ARC and the infundibular nucleus but, contrasting strikingly with the rat, the lemur exhibits marked labeling in the hypothalamic periventricular nucleus and the pars tuberalis of the pituitary gland, whereas no labeling is detectable in the VMN and the lateral hypothalamic area.

In the pituitary, GHS-R expression is found by reverse transcriptase polymerase chain reaction, in every tested GH-, GH/prolactin (PRL)- and PRL-secreting adenomas, in two out of three corticotroph, two out of four gonadotroph and one out of five non-secreting tumors but it is not observed in one thyroid-stimulating hormone-secreting adenoma (17). Qualitatively similar results were also reported in another study (18). Triple in-situ hybridization shows colocalization of GHS-R mRNA with GH and PRL mRNAs, conjointly or separately, in individual cells of somatotroph, mammosomatotroph, and lactotroph adenomas (17).

3.   EFFECTS OF PREPROGHRELIN-DERIVED PEPTIDES ON PITUITARY AND HYPOTHALAMIC SECRETION

Parallel to ghrelin characterization, a cDNA was isolated from a mouse stomach library encoding a protein named prepromotilin-related peptide (ppMTLRP) which shares sequence similarities with prepromotilin (6). Mouse and rat ppMTLRP sequences are identical and show 89% identity with human ghrelin. By analogy with promotilin, cleavage of proMTLRP into a 18 aminoacid endogenous processed peptide was proposed on the basis of a conserved dibasic motif in position 9-10 of its sequence (Figure 4-1). The GH releasing activities of rat ghrelin/MTLRP and of human ghrelin/MTRLP, compared with that of the 18 aminoacid form of the peptide, ghrelin18, are equivalent on superfused pituitary explants *in vitro* (19). In contrast, the short peptide devoid of Ser-3 N-octanoylation ghrelin18[-] is inactive. In freely moving animals, both rat and human ghrelins (10 µg, intravenously) stimulate GH release, whereas the same dose of ghrelin18 or of ghrelin18[-] is ineffective. In the rat, ghrelin (10 µg, intravenously)   does   not   modify   plasma   concentrations   of   PRL, adrenocorticotropic hormone (ACTH) and leptin. In contrast, in human subjects ghrelin readily increases prolactin and ACTH or cortisol secretion (20,21). Accordingly, MK-0677 stimulates, in a dose-dependent way, hormonal release from all tested somatotroph, mammosomatotroph, lactotroph and corticotroph adenomas (17). Rat ghrelin also stimulates GH and PRL release in the tilapia, *Oreochromis mossambicus* (22). However, in this species, KP-102 (D-ala-D-beta-Nal-ala-trp-D-phe-lys-NH(2)), a synthetic GHS, is only active on GH (23).

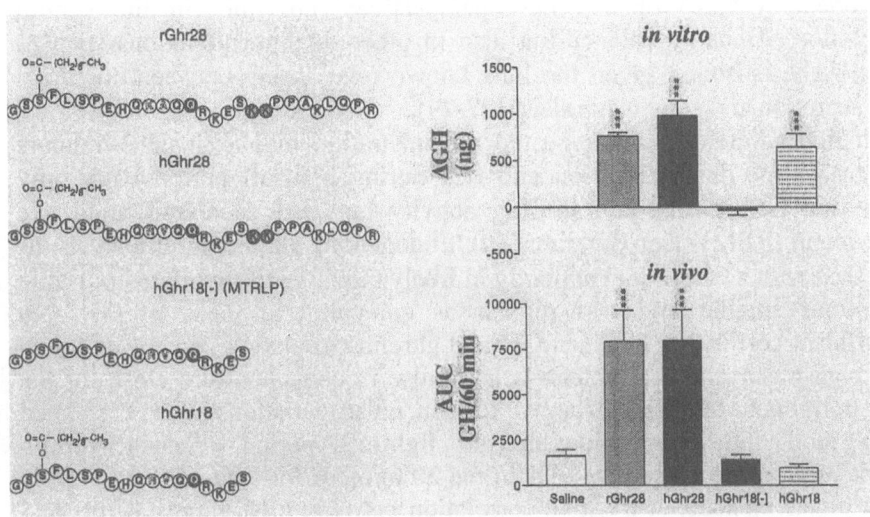

Figure 4-1. Effect of putative prepoghrelin derived peptides on GH release *in vitro* and *in vivo*. rGhr: rodent ghrelin; hGhr: human ghrelin; ***: p<0.001 vs saline [adapted from (19)]

Spontaneous GH secretory episodes in adult male rats are no longer observed within three hours after ghrelin treatment, but repeated administration of the peptide at 3-4 hour intervals results in similar GH responses. Activation of SRIF release by ether stress does not blunt the GH response to ghrelin. Since ether stress is likely to blunt GH secretion through activation of SRIF release from median eminence nerve terminals, this suggests that ghrelin acts in part by inhibiting endogenous SRIF, as further substantiated by its ability to decrease the amplitude of 25 mM K⁺-induced SRIF release from perifused hypothalamus (19). Passive immunization studies, using selective anti-SRIF and anti-GHRH antibodies, indicate that ghrelin is a functional antagonist of SRIF, but its GH-releasing activity at the pituitary level is not dependent on inhibiting endogenous SRIF release. On the other hand, the GH response to ghrelin requires an intact endogenous GHRH system (14). Finally, ghrelin has been reported to increase GHRH and NPY efflux from rat hypothalamic explants in long term (45 min) incubations, and, to a lesser extent, corticotropin-releasing hormone and vasopressin (24).

4.   GHRELIN, FOOD INTAKE AND SLEEP-WAKE RHYTHMICITY

The discovery that ghrelin injections increase food intake and adiposity in rodents (25,26) opened new perspectives, not only in the central regulation of energy balance but also in other rhythmic functions such as sleep/wake episodes, given the long known relationship between nocturnal GH secretion and slow wave sleep (27-29).

In the adult male rat, repeated administration of ghrelin at 3-4 hours intervals (one during light-on and two during light-off periods) not only increased GH release and feeding activity but also decreased rapid eye movement (REM) sleep duration (30). Endogenous plasma ghrelin levels, as assessed with a C-terminal antibody in freely moving rats, exhibited pulsatile variations smaller and less regular as compared to those of GH. No significant correlation between GH and ghrelin circulating levels was found although mean interpeak intervals and pulse frequencies were close for the two hormones. Most importantly, ghrelin pulse variations were correlated with food intake episodes in the light-off period. Plasma ghrelin concentrations decreased by 26% in the 20 minutes following the end of the food intake episode. A positive correlation between high ghrelin levels and active wakefulness occured during the first 3 hours of the dark period only (*ie*, a period in which the animals are actively eating).

Similar results are observed in the human species, both in terms of ghrelin effects on food intake (26) and relation between endogenous total ghrelin secretion and meal periodicity (31). In term of sleep regulation, one study indicates an increase in slow wave sleep after repeated ghrelin administration, every hour during the first 4 hours of the night, and a reduction in REM sleep thereafter (32) (Figure 4-2).

Both centrally (33) and systemically (34) administered ghrelin induce *c-fos* expression in a subpopulation of periventricular ARC neurons, a region where NPY cells are located. Moreover, centrally administered ghrelin increases hypothalamic NPY mRNA expression (35,36). Blocking NPY action by pretreatment with either NPY antiserum or specific NPY receptor antagonists (33,36) significantly interfere with ghrelin appetite-stimulating effect. However, GH-Releasing Peptide-2 still increases acutely appetite in knock-out mice lacking NPY and this effect is blocked by a melanocortin receptor agonist (37), thereby suggesting that AgRP and/or GABA which colocalize with NPY are probably involved in the orexigenic function of these neurons. Finally, blockade of gastric vagal afferents abolishes ghrelin-induced feeding, GH secretion, and *c-fos* activation in NPY- and GHRH-producing neurons (38). Thus, ghrelin receptors synthesized in vagal afferent neurons and transported to the afferent terminals are also involved in the central actions of ghrelin.

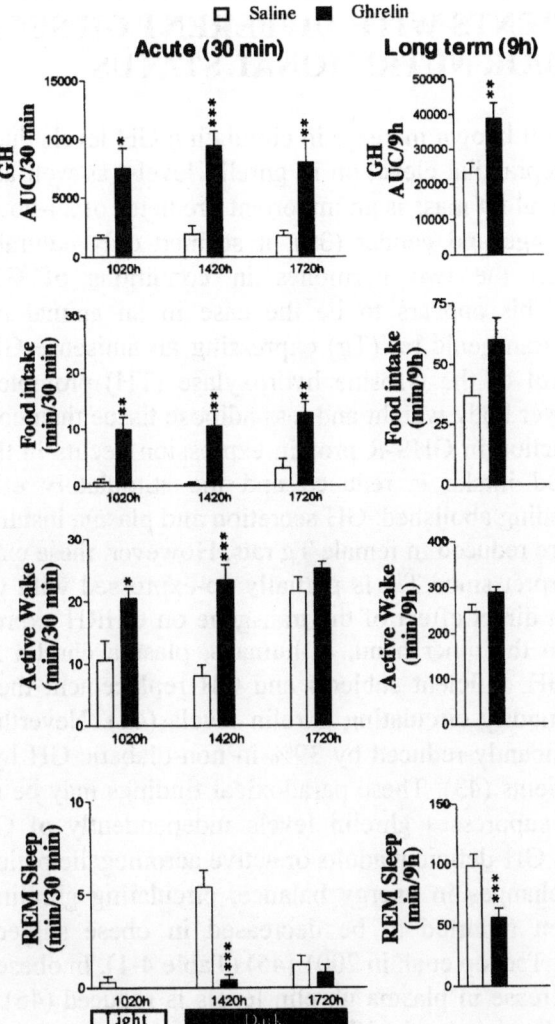

Figure 4-2. Comparison of the acute and long term effects of ghrelin injections on GH secretion, food intake and sleep wake patterns. *: p<0.05 vs saline; **: p<0.01 vs saline; ***: p<0.001 vs saline.

5.   ENDOGENOUS GHRELIN SECRETION IN PATIENTS WITH DIFFERENT GH SECRETORY AND/OR NUTRITIONAL STATUS

Given the well known increase in circulating GH levels during fasting in man and the preprandial elevation in ghrelin levels, as well as the fact that abdominal visceral fat mass is an important predictor of 24-hour GH release, independent of age and gender (39), it seemed only natural to predict a relation between the two hormones in conditions of GH hyper- or hyposecretion. This appears to be the case in an animal model of GH deficiency, the transgenic rat (Tg) expressing an antisense GHS-R mRNA under the control of the tyrosine hydroxylase (TH) promoter. These rats present with lower body weight and less adipose tissue than control rats and a selective reduction in GHS-R protein expression occurs in the ARC (40). Their daily food intake is reduced and the stimulatory effect of GHS treatment on feeding abolished. GH secretion and plasma insulin-like growth factor-I levels are reduced in female Tg rats. However, these experiments are difficult to interpret since TH is partially co-expressed with GHRH in the ARC (41) and a direct effect of the transgene on GHRH expression cannot be excluded. On the other hand, in humans, plasma ghrelin levels appear unchanged in GH deficient subjects and GH replacement therapy for one year does not modify circulating ghrelin levels (42). Nevertheless ghrelin levels are significantly reduced by 39% in non-diabetic GH hypersecretory acromegalic patients (43). These paradoxical findings may be related to the fact that SRIF suppresses ghrelin levels independently of GH status, in normal subjects, GH-deficient adults or active acromegalic patients (44).

In term of changes in energy balance, circulating ghrelin levels have consistently been reported to be decreased in obese subjects since the original study of Tschöp et al. in 2001 (45) (Table 4-1). In obese patients, the postprandial decrease in plasma ghrelin levels is reduced (46). Conversely, ghrelin plasma levels are significantly increased in anorexia (47,48) and bulimia nervosa (49). Interestingly, constitutionally thin women without body image disturbance or psychological disorders, with a low body mass index equivalent to anorexia nervosa patients but normal percentages of fat mass, do not display elevated plasma ghrelin concentrations (50). This confirms that circulating ghrelin levels depend on body fat mass. It remains to be demonstrated whether they are also directly influenced by feeding behavior and/or energy intake habits. Genetic factors may also be involved since a twin study indicated that plasma ghrelin concentration is a familial trait (51) and single nucleotide polymorphisms (SNPs) were evidenced in the preproghrelin gene (52,53). However, conflicting results have been obtained concerning the influence of these SNPs on weight regulation (54,55).

Finally, a noticeable exception to the lack of change in ghrelin levels in GH deficient patients is Prader Willy syndrome. In these patients, GH deficiency is associated with obesity and fasting circulating ghrelin levels are very elevated (56,57).

Table 4-1. Plasma ghrelin levels in patients with different GH and/or nutritional secretion status

GH status	Food intake	Circulating ghrelin levels	Ref.
Hypersecretion			
Acromegaly	not available	decreased	(43)
Anorexia Nervosa	decreased	increased	(47)
Bulimia Nervosa	binge eating	increased	(48, 49)
Hyposecretion			
GH deficient adults	not available	not changed	(42, 58)
Elderly subjects	decreased	decreased	(59)
Prader Willy Syndrome	increased	increased	(56, 57)
Obesity	increased	decreased	(45, 46, 56, 60, 61)

6.   CONCLUSION

Since its discovery in December 1999 as a GH secretagogue, the emphasis on ghrelin's physiological roles has shifted to its implication in energy balance regulation. Data concerning sleep regulation remain scarce. Ghrelin is the only peripheral hormone which stimulates appetite and fat storage; the former is congruent with its potent GH-secretagogue activity but the latter is rather puzzling in light of the well known lipolytic function of GH. This may suggest that different receptors or different locations (central and peripheral) of these receptors are involved in the GH-secretagogue and energy balance functions of ghrelin. Data obtained on circulating ghrelin levels in various physiological and pathological conditions have not yet permitted resolution of this paradox. However, it should be kept in mind that they were obtained with measurements of total (octanoylated and non-octanoylated) ghrelin immunoreactivity while the active forms represent less than 10% of total (7).

Post Scriptum: since the submission of this manuscript a previously uncharacterized group of ghrelin immunoreactive neurons was identified in the anterior periventricular hypothalamus of the rat using an antibody against the non octanoylated form of ghrelin (62). These neurons send efferents onto those producing NPY, AgRP, proopiomelanocortin products, and corticotropin-releasing hormone; thus representing a novel regulatory circuit controlling energy homeostasis. However, no connections between these neurons and GHRH or SRIF neurons were evidenced. Moreover, it was shown that a ghrelin-secreting human pancreatic tumor did not cause acromegaly, thereby questioning the involvement of peripheral ghrelin secretion in the control of GH secretion (63).

ACKNOWLEDGMENTS

This work was supported by INSERM ATC nutrition.

REFERENCES

1 Kojima M, Hosoda H, Date Y, et al. Ghrelin is a growth-hormone-releasing acylated peptide from stomach. Nature. 1999; 402:656-60.
2 Howard AD, Feighner SD, Cully DF, et al. 1996 A receptor in pituitary and hypothalamus that functions in growth hormone release. Science. 1996; 273:974-7.
3 Patchett AA, Nargund RP, Tata JR, et al. Design and biological activities of L-163,191 (MK-0677): a potent, orally active growth hormone secretagogue. Proc Natl Acad Sci USA. 1995; 92:7001-5.
4 Bowers CY, Momany F, Reynolds GA, et al. Structure-activity relationships of a synthetic pentapeptide that specifically releases growth hormone *in vitro*. Endocrinology. 1980; 106:663-7.
5 Horvath TL, Diano S, Sotonyi P, et al. Minireview: ghrelin and the regulation of energy balance--a hypothalamic perspective. Endocrinology. 2001; 142:4163-9.
6 Tomasetto C, Karam SM, Ribieras S, et al. Identification and characterization of a novel gastric peptide hormone: the motilin-related peptide. Gastroenterology. 2000; 119:395-405.
7 Hosoda H, Kojima M, Matsuo H, et al. Ghrelin and des-acyl ghrelin: two major forms of rat ghrelin peptide in gastrointestinal tissue. Biochem Biophys Res Commun. 2000; 279:909-13.
8 Lu S, Guan JL, Wang QP, et al. Immunocytochemical observation of ghrelin-containing neurons in the rat arcuate nucleus. Neurosci Lett. 2002; 321:157-60.
9 Korbonits M, Bustin SA, Kojima M, et al. The expression of the growth hormone secretagogue receptor ligand ghrelin in normal and abnormal human pituitary and other neuroendocrine tumors. J Clin Endocrinol Metab. 2001; 86:881-7.
10 van der Toorn FM, Janssen JA, de Herder WW, et al. Central ghrelin production does not substantially contribute to systemic ghrelin concentrations: a study in two subjects with active acromegaly. Eur J Endocrinol. 2002; 147:195-9.

11 Bennett PA, Thomas GB, Howard AD, et al. Hypothalamic growth hormone secretagogue-receptor (GHS-R) expression is regulated by growth hormone in the rat. Endocrinology. 1997; 138:4552-7.

12 Tannenbaum GS, Lapointe M, Beaudet A, et al. Expression of growth hormone secretagogue-receptors by growth hormone-releasing hormone neurons in the mediobasal hypothalamus. Endocrinology. 1998; 139:4420-3.

13 Willesen MG, Kristensen P, Romer J. Co-localization of growth hormone secretagogue receptor and NPY mRNA in the arcuate nucleus of the rat. Neuroendocrinology. 1999; 70:306-16.

14 Tannenbaum GS, Epelbaum J, Bowers CY. Interrelationship between the novel peptide ghrelin and somatostatin/growth hormone (GH)-releasing hormone in regulation of pulsatile GH secretion. Endocrinology. 2003; 144: 967-74.

15 Guan XM, Yu H, Palyha OC, et al. Distribution of mRNA encoding the growth hormone secretagogue receptor in brain and peripheral tissues. Brain Res Mol Brain Res. 1997; 48:23-9.

16 Mitchell V, Bouret S, Beauvillain JC, et al. Comparative distribution of mRNA encoding the growth hormone secretagogue-receptor (GHS-R) in Microcebus murinus (Primate, lemurian) and rat forebrain and pituitary. J Comp Neurol. 2001; 429:469-89.

17 Barlier A, Zamora AJ, Grino M, et al. Expression of functional growth hormone secretagogue receptors in human pituitary adenomas: polymerase chain reaction, triple in-situ hybridization and cell culture studies. J Neuroendocrinol. 1999; 11:491-502.

18 Korbonits M, Jacobs RA, Aylwin SJ, et al. Expression of the growth hormone secretagogue receptor in pituitary adenomas and other neuroendocrine tumors. J Clin Endocrinol Metab. 1998; 83:3624-30.

19 Tolle V, Zizzari P, Tomasetto C, et al. *In vivo* and *in vitro* effects of ghrelin/motilin-related peptide on growth hormone secretion in the rat. Neuroendocrinology. 2001; 73:54-61.

20 Takaya K, Ariyasu H, Kanamoto N, et al. Ghrelin strongly stimulates growth hormone release in humans. J Clin Endocrinol Metab. 2000; 85:4908-11.

21 Arvat E, Maccario M, Di Vito L, et al. Endocrine activities of ghrelin, a natural growth hormone secretagogue (GHS), in humans: comparison and interactions with hexarelin, a nonnatural peptidyl GHS, and GH-releasing hormone. J Clin Endocrinol Metab. 2001; 86:1169-74.

22 Riley LG, Hirano T, Grau EG. Rat ghrelin stimulates growth hormone and prolactin release in the tilapia, Oreochromis mossambicus. Zoolog Sci. 2002; 19:797-800.

23 Shepherd BS, Eckert SM, Parhar IS, et al. The hexapeptide KP-102 (D-ala-D-beta-Nal-ala-trp-D-phe-lys-NH(2)) stimulates growth hormone release in a cichlid fish (Ooreochromis mossambicus). J Endocrinol. 2000; 167:R7-10.

24 Wren AM, Small CJ, Fribbens CV, et al. The hypothalamic mechanisms of the hypophysiotropic action of ghrelin. Neuroendocrinology. 2002; 76:316-24.

25 Tschop M, Smiley DL, Heiman ML. Ghrelin induces adiposity in rodents. Nature. 2000; 407:908-13.

26 Wren AM, Seal LJ, Cohen MA, et al. Ghrelin enhances appetite and increases food intake in humans. J Clin Endocrinol Metab. 2001; 86:5992-5.

27 Sassin JF, Parker DC, Mace JW, et al. Human growth hormone release: relation to slow-wave sleep and sleep-walking cycles. Science. 1969; 165:513-5.

28 Kimura F, Tsai CW. Ultradian rhythm of growth hormone secretion and sleep in the adult male rat. J Physiol. 1984; 353:305-15.

29 Van Cauter E, Kerkhofs M, Caufriez A, et al. A quantitative estimation of growth hormone secretion in normal man: reproducibility and relation to sleep and time of day. J Clin Endocrinol Metab. 1992; 74:1441-50.

30 Tolle V, Bassant MH, Zizzari P, et al. Ultradian rhythmicity of ghrelin secretion in relation with GH, feeding behavior, and sleep-wake patterns in rats. Endocrinology. 2002; 143:1353-61.

31 Cummings DE, Purnell JQ, Frayo RS, et al. A preprandial rise in plasma ghrelin levels suggests a role in meal initiation in humans. Diabetes. 2001; 50:1714-9.

32 Weikel JC, Wichniak A, Ising M, et al. Ghrelin promotes slow-wave sleep in man. Am J Physiol Endocrinol Metab. 2002; E407-E415.

33 Nakazato M, Murakami N, Date Y, et al. A role for ghrelin in the central regulation of feeding. Nature. 2001; 409:194-8.

34 Hewson AK, Dickson SL. Systemic administration of ghrelin induces Fos and Egr-1 proteins in the hypothalamic arcuate nucleus of fasted and fed rats. J Neuroendocrinol. 2000; 12:1047-9.

35 Kamegai J, Tamura H, Shimizu T, et al. Chronic central infusion of ghrelin increases hypothalamic neuropeptide Y and Agouti-related protein mRNA levels and body weight in rats. Diabetes. 2001; 50:2438-43.

36 Shintani M, Ogawa Y, Ebihara K, et al. Ghrelin, an endogenous growth hormone secretagogue, is a novel orexigenic peptide that antagonizes leptin action through the activation of hypothalamic neuropeptide Y/Y1 receptor pathway. Diabetes. 2001; 50:227-32.

37 Tschöp M, Statnick MA, Suter TM, et al. GH-releasing peptide-2 increases fat mass in mice lacking NPY: indication for a crucial mediating role of hypothalamic agouti-related protein. Endocrinology. 2002; 143:558-68.

38 Date Y, Murakami N, Toshinai K, et al. The role of the gastric afferent vagal nerve in ghrelin-induced feeding and growth hormone secretion in rats. Gastroenterology. 2002; 123:1120-8.

39 Clasey JL, Weltman A, Patrie J, et al. Abdominal visceral fat and fasting insulin are important predictors of 24-hour GH release independent of age, gender, and other physiological factors. J Clin Endocrinol Metab. 2001; 86:3845-52.

40 Shuto Y, Shibasaki T, Otagiri A, et al. Hypothalamic growth hormone secretagogue receptor regulates growth hormone secretion, feeding, and adiposity. J Clin Invest. 2002; 109:1429-36.

41 Daikoku S, Kawano H, Noguchi M, et al. GRF neurons in the rat hypothalamus. Brain Res. 1986; 399:250-61.

42 Janssen JA, van der Toorn FM, Hofland LJ, et al. Systemic ghrelin levels in subjects with growth hormone deficiency are not modified by one year of growth hormone replacement therapy. Eur J Endocrinol. 2001; 145:711-6.

43 Cappiello V, Ronchi C, Morpurgo PS, et al. Circulating ghrelin levels in basal conditions and during glucose tolerance test in acromegalic patients. Eur J Endocrinol. 2002; 147:189-94.

44 Norrelund H, Hansen TK, H OR, et al. Ghrelin immunoreactivity in human plasma is suppressed by somatostatin. Clin Endocrinol. 2002; 57:539-46.

45 Tschop M, Weyer C, Tataranni PA, et al. Circulating ghrelin levels are decreased in human obesity. Diabetes. 2001; 50:707-9.

46 English PJ, Ghatei MA, Malik IA, et al. Food fails to suppress ghrelin levels in obese humans. J Clin Endocrinol Metab. 2002; 87:2984-7.

47 Ariyasu H, Takaya K, Tagami T, et al. Stomach is a major source of circulating ghrelin, and feeding state determines plasma ghrelin-like immunoreactivity levels in humans. J Clin Endocrinol Metab. 2001; 86:4753-8.

48 Otto B, Cuntz U, Fruehauf E, et al. Weight gain decreases elevated plasma ghrelin concentrations of patients with anorexia nervosa. Eur J Endocrinol. 2001; 145:669-73.

49 Tanaka M, Naruo T, Muranaga T, et al. Increased fasting plasma ghrelin levels in patients with bulimia nervosa. Eur J Endocrinol. 2002; 146:R1-3.

50 Tolle V, Kadem M, Bluet-Pajot MT, et al. Balance in ghrelin and leptin plasma levels in anorexia nervosa patients and constitutionally thin women. J Clin Endocrinol Metab. 2003; 88:109-16.

51 Ravussin E, Tschop M, Morales S, et al. Plasma ghrelin concentration and energy balance: overfeeding and negative energy balance studies in twins. J Clin Endocrinol Metab. 2001; 86:4547-51.

52 Ukkola O, Ravussin E, Jacobson P, et al. Mutations in the preproghrelin/ghrelin gene associated with obesity in humans. J Clin Endocrinol Metab. 2001; 86:3996-9.

53 Korbonits M, Gueorguiev M, O'Grady E, et al. A variation in the ghrelin gene increases weight and decreases insulin secretion in tall, obese children. J Clin Endocrinol Metab. 2002; 87:4005-8.

54 Hinney A, Hoch A, Geller F, et al. Ghrelin gene: identification of missense variants and a frameshift mutation in extremely obese children and adolescents and healthy normal weight students. J Clin Endocrinol Metab. 2002; 87:2716-9.

55 Ukkola O, Poykko S. Ghrelin, growth and obesity. Ann Med. 2002; 34:102-8.

56 Cummings DE, Clement K, Purnell JQ, et al. Elevated plasma ghrelin levels in Prader Willi syndrome. Nat Med. 2002; 8:643-4.

57 DelParigi A, Tschop M, Heiman ML, et al. High circulating ghrelin: a potential cause for hyperphagia and obesity in prader-willi syndrome. J Clin Endocrinol Metab. 2002; 87:5461-4.

58 Dall R, Kanaley J, Hansen TK, et al. Plasma ghrelin levels during exercise in healthy subjects and in growth hormone-deficient patients. Eur J Endocrinol. 2002; 147:65-70.

59 Rigamonti AE, Pincelli AI, Corra B, et al. Plasma ghrelin concentrations in elderly subjects: comparison with anorexic and obese patients. J Endocrinol. 2002; 175:R1-5.

60 Lindeman JH, Pijl H, Van Dielen FM, et al. Ghrelin and the hyposomatotropism of obesity. Obes Res. 2002; 10:1161-6.

61 Shiiya T, Nakazato M, Mizuta M, et al. Plasma ghrelin levels in lean and obese humans and the effect of glucose on ghrelin secretion. J Clin Endocrinol Metab. 2002; 87:240-4.

62. Cowley MA, Smith RG, Diano S, et al. The distribution and mechanism of action of ghrelin in the CNS demonstrates a novel hypothalamic circuit regulating energy homeostasis. Neuron. 2003; 37(4):649-61.

63. Corbetta S, Peracchi M, Cappiello V, et al. Circulating ghrelin levels in patients with pancreatic and gastrointestinal neuroendocrine tumors: identification of one pancreatic ghrelinoma. J Clin Endocrinol Metab. 2003; 88(7):3117-20.

Chapter 5

GROWTH HORMONE RELEASING ACTIVITY OF GHRELIN

Carlos Diéguez, Fernando Cordido, Vera Popovic[1] & Felipe F. Casanueva
Departments of Medicine and Physiology, University of Santiago de Compostela, Spain;
[1]Institute of Endocrinology, University of Belgrade, Belgrade, Serbia.

Abstract: The isolation of ghrelin is one of the most important breakthroughs in the understanding of the regulatory mechanisms involved in the neuroregulation of growth hormone (GH) secretion for several reasons. It gives definitive proof of the existence of a GH Secretagogue (GHS)/GHS-receptor signaling system in the control of GH secretion. Although for many years it was dogma that GH secretion by the anterior pituitary gland was the net result of the antagonistic actions of growth hormone releasing hormone (GHRH) and somatostatin, now a new physiological model of the regulation of GH secretion involving GHRH, somatostatin and ghrelin must be developed. It opens up the possibility of gaining a greater insight into the physiopathological mechanisms involved in the alterations of somatotroph cell function and somatic growth. Finally, it will allow the development of new agonist and antagonist compounds that may well be useful in the treatment of different disease states.

Key words: somatotroph, GH, Pit-1, GHRH, somatostatin

1.   INTRODUCTION

Growth hormone Secretagogues (GHS) are synthetic peptidyl and nonpeptidyl (MK-0677) molecules with strong, dose-dependent, and reproducible growth hormone (GH)-releasing activity even after oral administration. GHS release GH via actions on specific receptors (GHS-R) at the pituitary and, mainly, at the hypothalamic levels. GHS likely act as functional somatostatin antagonists and enhance the activity of GH-releasing hormone (GHRH)-secreting neurons (1). With the subsequent cloning of the specific GHS-R (2), and of the natural ligand for that receptor, named ghrelin (3), it is clear that GHS and the GHS-R are part of a new physiological system involved in GH regulation. GH secretion is mainly dependent on the interaction between GHRH, ghrelin and somatostatin (1).

In this chapter we will review recent data regarding the role of ghrelin on GH secretion, focusing on recent reports in which the effects of this newly discovered hormone have been assessed. Earlier works with synthetic GHS were reviewed previously (1,4-5).

2.   REGULATION OF GH SECRETION BY GHRELIN

Kojima and coworkers (3) first described a marked stimulatory effect of ghrelin *in vivo* using pentobarbital-anaesthetized rats. Since this experimental model is limited by the fact that pentobarbital inhibits hypothalamic somatostatin release, it was important to assess its effect in freely-moving rats, a model that allows all the regulatory mechanisms to be operating. Administration of ghrelin to normal freely-moving rats led to a marked increase in plasma GH levels in comparison to control untreated rats. A maximal stimulatory effect on plasma GH levels was observed following administration of 12 nmol/kg of ghrelin, the effect being similar to the one obtained with 60 nmol/kg in terms of both area under curves (AUC) and mean peak GH levels (Fig. 7-1). Interestingly, while at the dose of 3 nmol/kg GHRH and ghrelin exhibited a similar stimulatory effect, following administration of a dose of 12 nmol/kg, the effect of ghrelin was much greater than the same dose of GHRH in terms of both AUC and mean peak GH levels. However, the maximal stimulatory effect exerted by ghrelin was two to three times greater than GHRH in terms of both AUC and mean peak GH levels (6).The interest of this finding is based on the fact that it shows that *in vivo* ghrelin is a more potent GH releaser than GHRH, which is in contrast to their *in vitro* effects where the opposite was described (3). Nevertheless, it should be noted that it is well established that GHRH is needed for somatotroph cell proliferation and that GH synthesis and GH secretion elicited by most GH stimuli including synthetic compounds that activate the GHS-R is absent after administration of a GHRH-antagonist as well after passive immunization with anti-GHRH antiserum (4,5). Thus, it could be postulated that while GHRH may well play a pleiotropic role on somatotroph cell function, GH secretion will be dependent on the antagonistic effects exerted by ghrelin and somatostatin.

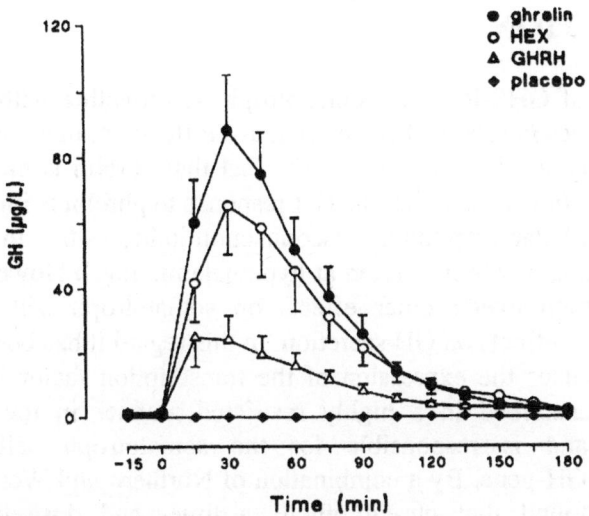

Figure 5-1. GH responses in normal subjects after the acute intravenous administration of ghrelin, hexarelin and GHRH (1.0 µg/kg) [from (15)]

To assess the physiological role of ghrelin on GH secretion, Shuto and coworkers (7) created transgenic rats expressing an antisense GHS-R mRNA under the control of the promoter for tyrosine hydroxylase, thus selectively attenuating GHS-R protein expression in the arcuate nucleus (ARC). GHRH-containing neurons in the ARC have been proposed to be one of the central targets of ghrelin in stimulating GH secretion, since many of these neurons coexpressed GHRH and GHS-R, and ghrelin administration leads to a marked increase in c-fos immunoreactivity (8). They found that pulsatile GH secretion was preserved in male rats with a functional knock-out of the GHS-R in the ARC. However, female transgenic rats showed lower baseline GH levels and fewer pulses than normal female rats. Furthermore, plasma insulin-growth factor-I (IGF-I) levels were also lower in female transgenic rats indicating that their altered GH secretion is of functional significance. Thus, it seems that in female rats, ghrelin plays a more important role than in males in the regulation of GH secretion by stimulating GHRH neurons through the GHS-R. Future studies assessing ghrelin, GHRH and somatostatin levels in portal blood vessels and their relationship to GH pulses should answer this question.

3.   GHRELIN AND SOMATOTROPH CELL FUNCTION

The presence of GHS-R in the somatotrophs (2) together with the fact that ghrelin increases *in vitro* GH secretion leave little doubt that ghrelin acts directly at pituitary level. Nevertheless, the fact that ghrelin is much more potent *in vivo* than *in vitro* and that the GH response to ghrelin is impaired in patients with hypothalamus-pituitary disconnection indicates that in terms of GH secretion its major role is exerted at hypothalamic level. However, it is possible that ghrelin exerts other effects on somatotroph cell function independently of its effects on GH secretion. In this regard it has been shown that ghrelin influences the expression of the transcription factor Pit-1 (9). This factor is transcribed in a highly restricted manner in the anterior pituitary gland and is responsible for the somatotroph cell-specific expression of the GH-gene. By a combination of Northern and Western blot analysis it was found that ghrelin elicits a time- and dose-dependent activation of Pit-1 expression in monolayer cultures of infant rat anterior pituitary cells. The effect was blocked by pretreatment with actinomycin-D but not by cicloheximide, suggesting that this action was due to direct transcriptional activation of Pit-1. Further assessment of the responsive regions of the Pit-1 promoter showed that the effect of ghrelin takes place in a sequence that contains two cyclic AMP response elements and that both of them are needed to induce the transcriptional activation of this gene. Although the transducing pathways that mediate this effect of ghrelin are not yet fully known, preliminary evidence suggest that it is dependent on protein kinase C, mitogen-activated protein kinase and protein kinase C activation. Taking into account the important role of Pit-1 in somatotroph cell-differentiation and cell-proliferation these data indicate that ghrelin, in addition to its effects on GH secretion, may play an important role in the physiological control and physiopathological alterations related to somatotroph cell function.

Although there is little doubt that the effects of GHRH and ghrelin on the somatotroph are mediated through different receptors the possible interaction among both receptors is of great interest and has been recently studied (10). It was reported that activation of GHS-R alone had no effect on cAMP production, coactivation of the GHS and GHRH receptors produced a cAMP response approximately twice that observed after activation of the GHRH receptor alone. This potentiated response is dose-dependent with respect to both GHRH and GHS, is dependent on the expression of both receptors, and was observed with a variety of peptide and nonpeptide GHS compounds as well as with ghrelin. Pharmacological inhibition of signaling molecules associated with GHS-R activation, including G-protein betagamma subunits, phospholipase C, and protein kinase C, had no effect on GHS potentiation of GHRH-induced cAMP production. Importantly, the potentiation appears to be selective for the GHRH receptor. Treatment of cells with the pharmacological agent forskolin elevated cAMP levels, but these levels were not further increased by GHS-R activation. Similarly, activation of two receptors homologous to the GHRH receptor, the vasoactive intestinal peptide and secretin receptors, increased cAMP levels, but these levels were not further increased by GHS-R activation. Based on these findings, it is possible that direct interactions between the GHRH and GHS-R may explain the observed effects on signal transduction.

The source of ghrelin involved in its direct effects at the pituitary level is still unsettled. Since relatively large concentrations of ghrelin are present in the circulating blood, mostly synthesized in the stomach, some authors suggest that stomach-derived ghrelin could influence somatotroph cell function. However, the fact that it is also synthesized in the pituitary open up the possibility of a paracrine effect (11).

4. EFFECTS OF GHRELIN AT THE HYPOTHALAMIC LEVEL

The wide distribution of GHS-R in the hypothalamus suggested that neurons present in this area could be a target for ghrelin. It was found that ghrelin administration led to c-fos activation in different subsets of hypothalamic neurons notably in the ARC (8). Since this nucleus plays an important role in the regulation of food intake (12) and GH secretion, it is not surprising that some of the more important biological actions of ghrelin are related to body weight homeostasis and the tuning of the hypothalamic-GH axis.

In order to characterize the hypothalamic action of ghrelin on neuropeptide gene expression different groups have reported the effect of ghrelin administration on neuropeptide Y (NPY) mRNA levels. Ghrelin administration to ad libitum fed rats led to a clear-cut increase in agouti-related peptide (AgRP) and NPY mRNA contents in the ARC (12). This effect appears to be quite specific as melanin-concentrating hormone and prepro-orexin in the lateral hypothalamus, GHRH in the ARC and somatostatin mRNA levels in the periventricular nucleus were unchanged after treatment with ghrelin(13). Since NPY plays an inhibitory role on GH secretion, it is likely that the stimulatory effect of ghrelin on AgRP/NPY neurons is linked to its orexigenic effects rather than to its actions in the somatotropic axis. Nevertheless, at first glance this data was somewhat surprinsing because there is evidence indicating that, at least from a functional point of view, somatostatin mediates some of the effects of ghrelin on GH secretion. Taking into account that GH has been found to markedly affect somatostatin and GHRH mRNA levels and that their neurons express GH receptors it was possible that the ghrelin-induced increase in circulating GH levels could be masking their effects on these two neuropeptides. To address this issue Seoane and collaborators (13) assessed the effects of ghrelin in dwarf rats which do not secrete GH and therefore any direct effect of ghrelin could be uncovered. Using this experimental approach, it was found that while, similar to intact rats, ghrelin failed to modify GHRH mRNA content in the ARC, it induced a clear-cut stimulatory effect on somatostatin mRNA content in the periventricular nucleus through a non-GHRH dependent mechanism.

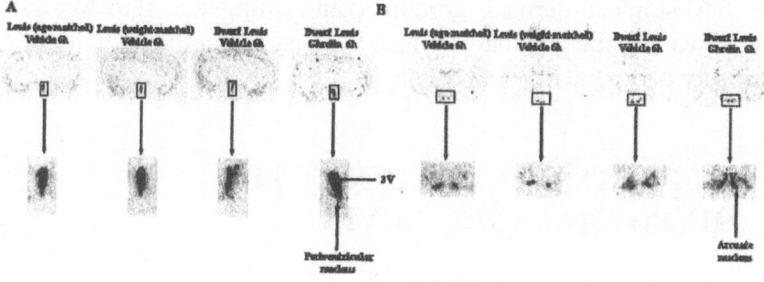

Figure 5-2. Autoradiographic images of representative brain coronal sections in either dwarf or age-matched Lewis rats treated with vehicle or ghrelin for 6 hours. Left panel shows somatostatin mRNA levels in the PeN. Right panel shows GHRH mRNA levels in the arcuate. The data shows that ghrelin is able to stimulate somatostatin mRNA levels in dwarf rats while it does not affect GHRH mRNA levels.

The relevance of the hypothalamic loci of action of ghrelin on GH secretion was recently addressed *in vivo* in rats with destruction of the ARC (14). Intravenous (iv) administration of ghrelin (10 µg/kg) increased plasma GH levels significantly in the normal adult male rats during a GH through period of pulsatile GH secretion, while iv injection of ghrelin in monosodium glutamate (MSG)-treated rats resulted in a markedly attenuated GH response. When rat ghrelin was administered intracerebroventricular (icv), plasma GH levels were increased comparably in normal control and MSG-treated rats. However, the GH release after icv injection of ghrelin was markedly diminished compared with that after iv administration of a small amount of ghrelin in normal control rats (icv: 10.0 µg/rat, iv: approximately 4.0 µg/rat), indicating that the GH-releasing activity of exogenous ghrelin is route-dependent and at least in part via hypothalamic ARC. The icv administration of 1.0 µg of ghrelin significantly increased 4-hour food intake in normal control, whereas the peptide did not increase food intake in MSG-treated rats, indicating that the feeding response to ghrelin requires intact ARC. Taken together, the primary action of ghrelin on appetite control and GH releasing activity was postulated to be via the ARC and to be mediated by another type of GHS-R besides GHS-R type 1a. Assessment of GH and food intake responses to ghrelin in GHS-R knock-out animals should clarify this question.

5. GHRELIN-INDUCED GH SECRETION IN HUMANS

There is now a general agreement that ghrelin elicits a marked increase in plasma GH levels in humans (15-17). The effect of the natural GHS ghrelin on GH secretion in humans showed that the GH response to ghrelin was clearly higher than the one recorded after GHRH and even significantly higher than after hexarelin (HEX), synthetic GHS. Ghrelin administration also induced an increase in prolactin, adrenocorticotropin hormone (ACTH), and cortisol levels; these responses were higher than those elicited by HEX. A significant increase in aldosterone levels was recorded after ghrelin but not after HEX administration (15). The co-administration of ghrelin and GHRH had a real synergistic effect on GH secretion in normal human subjects. As GH-Releasing Peptide (GHRP)-6 plus GHRH is a very potent GH releaser, attempts to verify whether this synergistic action also occurred for ghrelin were undertaken. In fact, ghrelin and the GHS HEX showed a strong potentiation of their GH secretory capability when injected together in humans, and more strikingly when ghrelin was employed at very low doses (15). This peculiar activity occurs due to a simultaneous ghrelin activation of pituitary and hypothalamic structures. Homologous and heterologous desensitization occurs after the administration of different GHS at high dosages, and could also occur when using ghrelin. As ghrelin is tonically released into the plasma from the stomach, it was of foremost importance to clarify whether this situation partially desensitizes central structures regulating GH secretion. In this regard recent studies have shown that ghrelin-induced GH secretion is resistant to homologous desensitization (18).

Taking into account that ghrelin plays an important role in energy homeostasis and that alterations in metabolic status markedly influence GH secretion (19), it was of great interest to assess circulating levels of ghrelin and GH in fed and fasted subjects. Unlike rodents, short-term fasting markedly increase spontaneous GH secretion. Despite intensive efforts by many different research groups the mechanisms responsible for fasting-driven increase in GH pulsatility in humans are yet unknown since it is appears that it is unrelated to changes in somatostatin or GHRH tone. Interestingly, fasting in humans leads to marked diurnal rhythm in circulating ghrelin levels that was associated to increase pulsatile GH secretion. Although intervention studies are needed, these data suggest that fasting-induced GH secretion could be driven by increased ghrelin secretion. (20).

In addition to the effect of fasting the possible influence of lipid-heparin infusion, hyperglycaemia and arginine on GH response to ghrelin have been recently reported (21). It is well known that free fatty acids (FFA) play an important physiological inhibitory role on GH secretion. (22) Administration of a lipid-heparin infusion that increases circulating levels of FFA, which leads to a partial decrease in ghrelin induced GH secretion. Similar findings were observed following the administration of an oral glucose load. Although the results were somewhat expected since similar findings were reported previously with synthetic GHS, this data contrasts clearly with that obtained with GHRH where an almost complete blockade of GH secretion was observed in the same experimental situation (21). Taken together this data suggests that the GH response to ghrelin is influenced to a lesser degree than to GHRH in relation to changes in metabolic status.

One of the major features of the regulation of GH secretion is that it is influenced by many central and peripheral signals. Thus, it was of interest to clarify their influence on somatotroph responsiveness following ghrelin administration.

Since somatostatin is the main inhibitory factor of GH secretion the attention of some groups focused on assessing its influence on both circulating ghrelin levels and GH responses to ghrelin. As expected somatostatin infusion inhibited GH responses to ghrelin though the effect was only partial (23,24). Interestingly somatostatin, as well as cortistatin-14 (23,24), inhibited circulating ghrelin levels indicating that the antagonistic effect of both hormones can take place at two different levels, e.g., inhibiting ghrelin release by a direct effect at the stomach and at the somatotroph cell.

It has been known for many years that cholinergic muscarinic receptors and arginine play a facilitatory role on GH secretion in humans. The effect of arginine and the cholinergic agonist pyridostigmine on GH secretion is synergistic with that elicited by GHRH. In contrast, recent data have shown that cholinergic muscarinic receptors or arginine administration did not modify GH responses to ghrelin (21,25). Since there is a wealth of indirect evidence that indicates that the effect of arginine and cholinergic pathways is mediated by inhibiting somatostatin tone this data supports the hypothesis that ghrelin may well be acting as a functional somatostatin antagonist in terms of GH secretion.

6. CLINICAL IMPLICATIONS

The most important GH-related clinical application of GHS presently is its use as a diagnostic tool in suspected GH deficiency (GHD). GHS when administered together with GHRH exert a synergistic action on GH secretion and that combined administration is the most potent GH releaser to date. Clinical studies have demonstrated that the GHS+GHRH administration may be considered a suitable test of GH reserve in humans, as the GH secretion so elicited is not altered, at least to a large degree, by gender, adiposity, or age. The combined administration of GHRH plus GHS is able to discriminate between healthy subjects and patients with adult GHD, suggesting a considerable utility in the clinical setting (26,27). Although there is not data at present related to the use of ghrelin in patients with adult onset GHD, it is expected that it should be at least as appropriate as GHRP-6. Nevertheless the natural GHS, ghrelin has already been used in adult patients with isolated childhood-onset GHD (28). It was found that ghrelin is one of the most powerful provocative stimuli of GH secretion, even in those patients with isolated severe GHD. In this condition, however, the somatotroph response is markedly reduced while the lactotroph and corticotroph responsiveness to ghrelin is fully preserved, indicating that this endocrine activity is fully independent of mechanisms underlying the GH-releasing effect. These results do not support the hypothesis that impaired ghrelin responsiveness is a major cause of isolated GHD but suggest that ghrelin might represent a reliable provocative test to evaluate the maximal secretory capacity provided that appropriate cut-off limits are assumed. The combined GHS+GHRH administration may be employed as a diagnostic test in different clinical situations.

7. CONCLUDING REMARKS

Ghrelin is a 28-amino-acid peptide, predominantly produced by the stomach, showing a unique structure with an n-octanoyl ester at its third serine residue, which is essential for its potent stimulatory activity on somatotroph secretion. In fact, it has been demonstrated that ghrelin specifically stimulates GH secretion from both rat pituitary cells in culture and rats *in vivo*. It displays strong GH-releasing activity mediated by the hypothalamus-pituitary GHS-R that was found to be specific for a family of synthetic, orally active GHS. The discovery of ghrelin brings us to a new understanding of the regulation of GH secretion. However, ghrelin is much more than simply a natural GHS. It also acts on other central and peripheral receptors and exhibits other actions, including stimulation of lactotroph and corticotroph secretion, orexigenic, gastroenteropancreatic functions, and metabolic, cardiovascular and antiproliferative effects. So the activity of GHS is not fully specific for GH. Their slight prolactin-releasing activity probably comes from direct pituitary action. In physiological conditions, the ACTH-releasing activity of GHS is dependent on central actions; a direct action on GHS-R in pituitary ACTH-secreting tumors likely explains the peculiar ACTH and cortisol hyperresponsiveness to GHS in Cushing disease. GHS have specific receptor subtypes in other central and peripheral endocrine and nonendocrine tissues mediating GH-independent biologic activities. GHS influence sleep patterns, stimulate food intake, and have cardiovascular activities. GHS have specific binding in normal and neoplastic follicular-derived human thyroid tissue and inhibit the proliferation of follicular-derived neoplastic cell lines. Knowledge of the whole spectrum of biologic activities of ghrelin will provide new understanding of some critical aspects of neuroscience, metabolism and internal medicine. In fact, GHS were born more than 20 years ago as synthetic molecules, eliciting the hope that orally active GHS could be used to treat GHD as an alternative to recombinant human GH. However, the dream did not become reality and the usefulness of GHS as an anabolic anti-aging intervention restoring the GH/IGF-I axis in somatopause is still unclear. Instead, we now face the theoretic possibility that GHS analogues acting as agonists or antagonists could become candidate drugs for the treatment of pathophysiologic conditions in internal medicine totally unrelated to disorders of GH secretion (29).

ACKNOWLEDGEMENTS

This work was supported by grants from the DGICYT, the FISss and the Xunta de Galicia.

REFERENCES

1. Ghigo E, Boghen M, Casanueva FF, Dieguez C. Growth hormone secretagogues. Basic findings and clinical implications. Holland: Elsevier. 1999; 1-320.
2. Howard HD, Feighner SD, Cully DF, et al. A receptor in pituitary and hypothalamus that functions in growth hormone release. Science. 1996; 273:974-6.
3. Kojima M, Hosoda H, Data Y, Nakazato M, Matsuo H, Kangawa K. Ghrelin is a growth hormone-releasing acylated peptide from stomach. Nature. 1999; 402:656-8.
4. Dieguez C, Casanueva FF. Ghrelin: a step forward in the understanding of somatotroph cell function and growth regulation. Eur J Endocrinol. 2000; 142:913-7.
5. Dieguez C, Casanueva FF. Influence of metabolic substrates and obesity on growth hormone secretion. Trends Endocrinol Metab. 1995; 6:55-9.

6. Seoane LM, Tovar S, Baldelli R, et al. Ghrelin elicits a marked stimulatory effect on GH secretion in freely-moving rats. Eur J Endocrinol. 2000; 143:R7-9.
7. Shuto Y, Shibasaki T, Otagiri A, et al. Hypothalamic growth hormone secretagogue receptor regulates growth hormone secretion, feeding and adiposity. J Clin Invest. 2002; 109:1429-36.
8. Dickson SL. Ghrelin: a newly discovered hormone. J Neuroendocrinol. 2002; 14:83-4.
9. Garcia A, Alvarez CV, Smith RG, Dieguez C. Regulation of Pit-1 expression by ghrelin and GHRP-6 through the GH secretagogue receptor. Mol Endocrinol. 2001; 15:1484-95.
10. Cunha SR, Mayo KE. Ghrelin and growth hormone (GH) secretagogues potentiate GH-releasing hormone (GHRH)-induced cyclic adenosine 3',5'-monophosphate production in cells expressing transfected GHRH and GH secretagogue receptors. Endocrinology. 2002; 143:4570-82.
11. Korbonits M, Kojima M, Kangawa K, Grossman AB. Presence of ghrelin in normal and adenomatous human pituitary. Endocrine. 2001; 14:101-4.
12. Muccioli G, Tschop M, Papotti M, Deghenghi R, Heiman M, Ghigo E. Neuroendocrine and peripheral activities of ghrelin: implications in metabolism and obesity. Eur J Pharmacol. 2002; 440:235-54.
13. Seoane LM, Lopez M, Tovar S, Casanueva FF, Senaris R, Dieguez C. Agouti-related Peptide, neuropeptide Y and somatostatin-producing neurons are targets for ghrelin actions in the rat hypothalamus. Endocrinology. 2003; 144:544-51.
14. Tamura H, Kamegai J, Shimizu T, Ishii S, Sugihara H, Oikawa S. Ghrelin stimulates GH but not food intake in arcuate nucleus ablated rats. Endocrinology. 2002; 143:3268-75.
15. Arvat E, Maccario M, Di Vito L, et al. Endocrine activities of ghrelin, a natural growth hormone secretagogue (GHS), in humans: comparison and interaction with hexarelin, a nonnatural peptidyl GHS, and GH-releasing hormone. J Clin Endocrinol Metab. 2001; 86:1169-74.
16. Takaya K, Ariyasu H, Kanamoto N, et al. Ghrelin strongly stimulates growth hormone release in humans. J Clin Endocrinol Metab. 2000; 85:4908-11.
17. Peino R, Baldelli R, Rodriguez-Garcia J, et al. Ghrelin-induced growth hormone secretion in humans. Eur J Endocrinol. 2000; 143:R11-4.
18. Micic D, Macut D, Sumarac-Dumanovic M, et al. Ghrelin-induced GH secretion in normal subjects is partially resistant to homologous desensitization by GH-releasing peptide-6. Eur J Endocrinol. 2002; 147:761-6.
19. Cordido F, Peñalva A, Dieguez C, Casanueva FF. Massive growth hormone discharge in obese subjects after the combined administration of GH-releasing hormone and GHRP-6: evidence for a marked somatotroph secretory capability in obesity. J Clin Endocrinol Metab. 1993; 76:819-23.
20. Muller AF, Lamberts SW, Janssen JA, et al. Ghrelin drives GH secretion during fasting in man. Eur J Endocrinol. 2002; 146:203-7.
21. Broglio F, Benso A, Gottero C, et al. Effects of glucose, free fatty acids or arginine load on the GH-releasing activity of ghrelin in humans. Clin Endocrinol (Oxf). 2002; 57:265-71.
22. Cordido F, Fernandez T, Martinez T, Peinó R, Dieguez C, Casanueva FF. Effect of acute pharmacological reduction of plasma free fatty acids on GHRH-induced GH secretion in obese adults with and without hypopituitarism. J Clin Endocrinol Metab. 1998; 83:4350-4.
23. Di Vito L, Broglio F, Benso A, et al. The GH-releasing effect of ghrelin, a natural GH secretagogue, is only blunted by the infusion of exogenous somatostatin in humans. Clin Endocrinol (Oxf). 2002; 56:643-8.

24. Broglio F, Koetsveld Pv P, Benso A, et al. Ghrelin secretion is inhibited by either somatostatin or cortistatin in humans. J Clin Endocrinol Metab. 2002; 87:4829-32.
25. Broglio F, Gottero C, Benso A, et al. Acetylcholine does not play a major role in mediating the endocrine responses to ghrelin, a natural ligand of the GH secretagogue receptor, in humans. Clin Endocrinol (Oxf). 2003; 58:92-8.
26. Popovic V, Pekic S, Golubic I, Doknic M, Dieguez C, Casanueva FF. The impact of cranial irradiation on GH responsiveness to GHRH plus GHRP-6. J Clin Endocrinol Metab. 2002; 87:2095-9.
27. Popovic V, Leal A, Micic D, et al. GH-releasing hormone and GH-releasing peptide-6 for diagnostic testing in GH-deficient adults. Lancet. 2000; 356:1137-42.
28. Aimaretti G, Baffoni C, Broglio F, et al. Endocrine responses to ghrelin in adult patients with isolated childhood-onset growth hormone deficiency. Clin Endocrinol (Oxf). 2002; 56:765-71.
29. Broglio F, Arvat E, Benso A, et al. Ghrelin: endocrine and non-endocrine actions. J Pediatr Endocrinol Metab. 2002; 5:219-27.

Chapter 6

NON-GROWTH HORMONE ENDOCRINE ACTIONS OF GHRELIN

Márta Korbonits & Ashley B. Grossman
Department of Endocrinology, St. Bartholomew's Hospital, London, EC1A 7BE, UK

Abstract: Apart from the growth hormone (GH) releasing effect of ghrelin, it also interacts with other hormonal systems, such as the hypothalamo-pituitary-adrenal (HPA) axis, prolactin secretion, the thyroid axis as well as the gonadal axis. Ghrelin and its analogues stimulate the HPA axis independent of the pituitary, via the hypothalamus, involving both corticotrophin-releasing hormone and arginine-vasopressin stimulation. However, the GH secretagogue receptor is pathologically expressed in some pituitary tumors, accounting for especially high adrenocortocotropic hormone (ACTH) and cortisol responses to ghrelin and GH secretagogues seen in patients with Cushing's disease. Ghrelin and its receptor are expressed both in the normal pituitary and adrenal gland, but no direct effect on ACTH/glucocorticoids has been observed. Ghrelin stimulates prolactin release most probably from the somato-mammotroph cells of the pituitary gland. The effects of ghrelin or its analogues on the pituitary regulation of the thyroid and gonadal axes seem to be minimal, although effects within the gonads and thyroid remain to be explored in more detail. Ghrelin and its receptor have been identified in various neuroendocrine tumors. Recent data identifying ghrelin in the placenta, and alterations in ghrelin levels in patients with polycystic ovarian disease, are worthy of further study.

Key words: ACTH, cortisol, thyroid, gonadal axis

1.   INTRODUCTION

The first recognized endocrine effect of the Growth Hormone Releasing Peptide (GHRP)-6 was naturally on growth hormone (GH) release. However, effects on other endocrine hormones were subsequently observed, while the later-identified natural analogue ghrelin turned out to be equally non-specific for GH release. In the current chapter we summarize the data available on the non-GH-related endocrine effects of ghrelin, and its synthetic analogues, on the hypothalamo-pituitary-adrenal (HPA), thyroid and gonadal axes.

2.   HYPOTHALAMO-PITUITARY-ADRENAL AXIS

2.1   The effect of ghrelin and GH secretagogues on adrenocorticotropic hormone and cortisol release

GH secretagogues (GHS) have been shown to activate the HPA axis in both animal and human studies (2-6). Originally, the effect on the HPA axis was not seen, as *in vitro* studies in pituitary culture do not show effects of the secretagogues on adrenocorticotropic hormone (ACTH) release, while the first animal studies with anesthetized animals failed to reveal any effect on the HPA axis (7,8). However, *in vivo* studies with unanesthetized animals established clear changes in both ACTH and cortisol in response to peptide or non-peptide secretagogues (5,9). In the human, activation of the pituitary-adrenal axis has been reported after intravenous (iv) administration of different synthetic GHS (Figure 6-1) (4,10-13), but not after oral administration of GHRP-6 (14). The ACTH and cortisol-releasing effects of GHRP-2 and hexarelin are comparable to each other, and also to that of human-sequence corticotrophin releasing hormone (h-CRH) (15). It has also been suggested that the deterioration in diabetic control and hyperlipidaemia seen in diabetic Zucker rats in response to GHS is probably due to their effect on the HPA axis (16).

The effects of the natural GHS receptor (GHS-R) ligand ghrelin on the HPA axis were also stimulatory, similar to the effects of the synthetic GHS. Previously, it had been considered that these changes might be non-specific effects of the synthetic peptide and non-peptide analogues. However, these changes have been reproducible with ghrelin. In human subjects, the effect of ghrelin on ACTH and cortisol release was higher after 1.0 µg/kg (equivalent to 3×10^{-10} mol/kg) iv ghrelin (peak [mean±standard error of the mean] ACTH: 20.3±4.5 pmol/l; cortisol: 503.2±43.5 nmol/l) than after 1.0 µg/kg (equivalent to 9×10^{-8} mol/kg) iv hexarelin (ACTH: 10.2±2.5 pmol/l; cortisol: 394.8±30 nmol/l) (17). In this study, stimulation of aldosterone secretion was also observed, probably due to the higher ACTH levels achieved by the ghrelin injection (Figure 6-2).

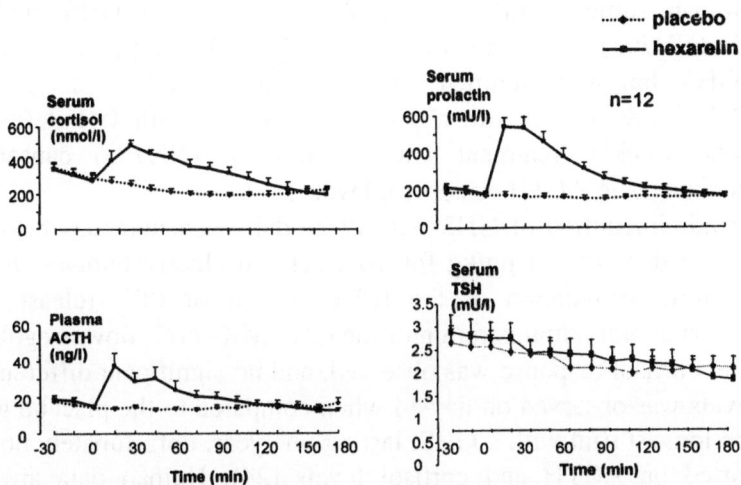

Figure 6-1 The effect of 2.0 µg/kg iv hexarelin in healthy volunteers. TSH: thyroid stimulating hormone. [adapted from (18)]

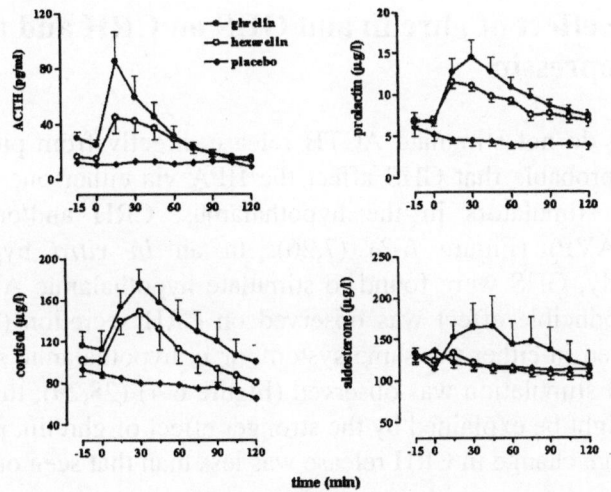

Figure 6-2 The effect of 1.0 µg/kg iv ghrelin compared to hexarelin in healthy male volunteers on ACTH, cortisol, prolactin (PRL) and aldosterone release. [adapted from (17)]

The GHS ipamorelin had been described as "the first selective GHS" following an investigation of its effects in swine (19). It still had an ACTH and cortisol-releasing activity compared to saline, but this was not significantly different from the ACTH and cortisol-releasing activity of GH releasing hormone (GHRH), the most selective GH-releasing agent. In healthy human subjects, the non-peptide synthetic oral GHS analogue tabimorelin (NN703) has been shown to have GH-releasing potency similar to other GHS, but with minimal effects on ACTH and prolactin (PRL) release (20). More recently, in a group of subjects with GH deficiency (GHD), one week's treatment with tabimorelin (NN703) caused no significant changes in ACTH or cortisol levels (21).

Longer administration of GHS with twice-daily subcutaneous hexarelin injections at a dose of 1.5 µg/kg for 16 weeks in elderly humans did not cause long-term stimulation of the HPA axis or of PRL release (22). Similarly, during oral administration of the GHS MK-0677, down-regulation of the initial cortisol response was observed, and no significant difference in cortisol levels was observed on day 14 when compared to the placebo group (23). In the longest trial with a GHS, lasting one year, unfortunately no data were reported on ACTH and cortisol levels (24). Human data are also available on the effect of somatostatin (SRIF) on ACTH and cortisol levels following GHS administration: it appears that the cortisol and ACTH-releasing effect of GHS is not abolished by SRIF infusion (25).

2.2 The effect of ghrelin and GHS on CRH and arginine-vasopressin

Since GHS do not stimulate ACTH release directly from pituitary cell cultures, it is probable that GHS affect the HPA via either one of the two major ACTH stimulators in the hypothalamus, CRH and/or arginine-vasopressin (AVP) (Figure 6-3) (7,26). In an *in vitro* hypothalamic incubation study, GHS were found to stimulate hypothalamic AVP release while no reproducible effect was observed on CRH secretion (27). When ghrelin was used in either the same system, or in hypothalamic slices, both AVP and CRH stimulation was observed (Figure 6-4) (28,29); this apparent discrepancy might be explained by the stronger effect of ghrelin, particularly as the percentage change in CRH release was less than that seen on AVP.

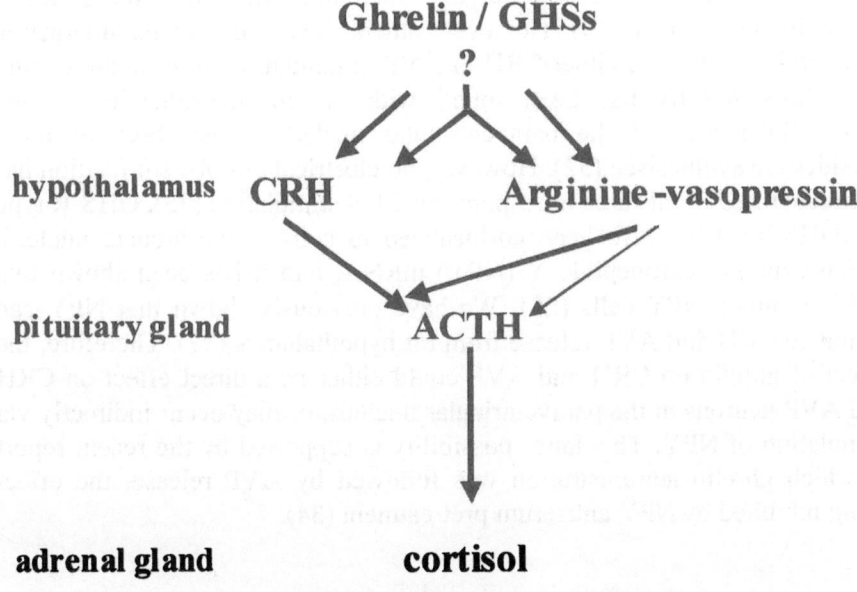

Figure 6-3 Regulation of the HPA axis by ghrelin

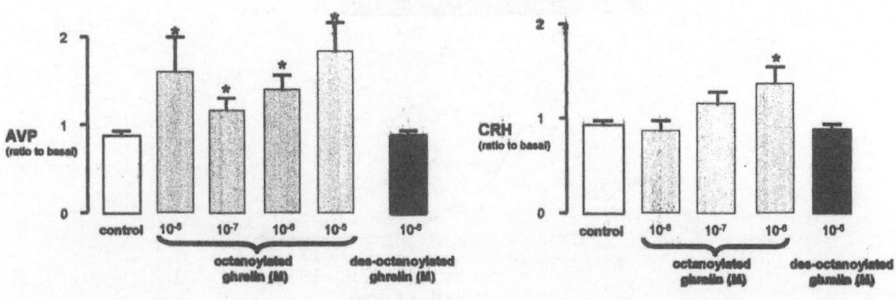

Figure 6-4 The effect of ghrelin on CRH and AVP release from rat hypothalamus *in vitro*; * p<0.05 vs control. [adapted from (29)]

An *in vivo* study in rats found that the co-administration of GHRP-6 and CRH caused similar ACTH elevations compared to CRH on its own, while the co-administration of GHRP-6 and AVP had a bigger effect on ACTH release, suggesting that GHRP synergizes with AVP in the release of ACTH, possibly by the stimulation of endogenous CRH (5). On the contrary, human data, using the synthetic analogue hexarelin, CRH and desmopressin injections, suggested a primary effect on AVP: the co-administration of hexarelin and CRH caused a higher ACTH and cortisol release than the effect of hexarelin and desmopressin, which was different to the effect of hexarelin on its own (30). However, others have reported no interaction between hexarelin and either CRH or AVP in humans (31). A small amount of GHS-R mRNA has been found with *in situ* hybridization in the parvocellular area of the paraventricular nucleus, where both of these peptides are synthesised (32). However, no electrical or c-*fos* stimulation has been identified in this area in response to GHS stimulation (33). GHS-R type 1a (GHS-R1a) has also been co-localized to cells in the arcuate nucleus, which express neuropeptide Y (NPY) mRNA, and it has been shown that GHS stimulate NPY cells (28). We have previously shown that NPY can stimulate CRH and AVP release from rat hypothalamus (27). Therefore, the effect of ghrelin on CRH and AVP could either be a direct effect on CRH and AVP neurons in the paraventricular nucleus, or may occur indirectly via stimulation of NPY. This latter possibility is supported by the recent report in which ghrelin administration was followed by AVP release, the effect being inhibited by NPY antiserum pretreatment (34).

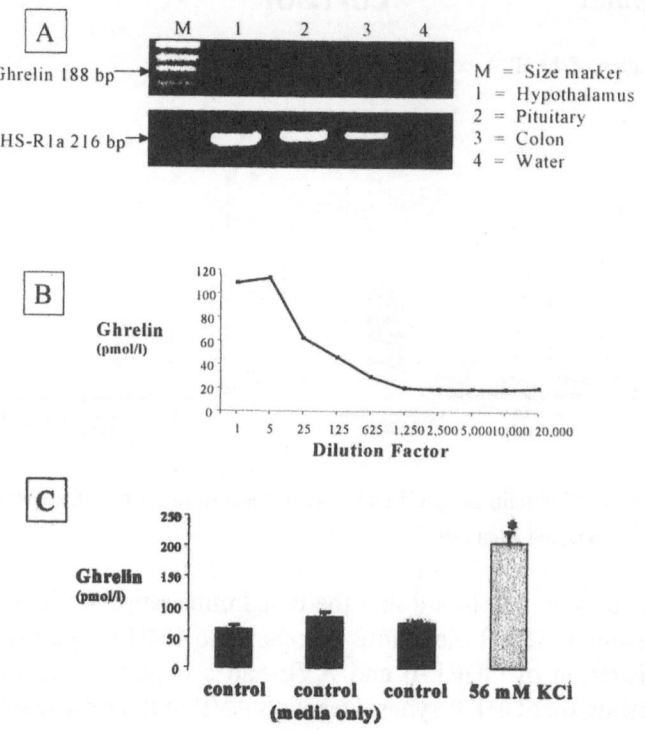

Figure 6-5. (A) Ghrelin and GHS-R1a mRNA expression in rat hypothalamus; (B) ghrelin content of rat hypothalamic tissue in serial dilutions; (C) ghrelin release from rat hypothalamus *in vitro.* * p<0.05 vs control [adapted from (29)]

Ghrelin is produced locally in the hypothalamus (Figure 6-5), therefore both locally produced ghrelin as well as ghrelin from the circulation can affect hypothalamic hormone systems.

2.3   Direct effects on the adrenal gland

The possibility of direct effects of GHS and ghrelin on the adrenal gland, and the expression of the GHS-R in the adrenal, has been studied by several groups. Ghrelin and GHS-R mRNA expression was found in the human adrenal tissue (35). No ghrelin expression was identified in rat adrenal tissue (36), but the active form of the GHS-R, GHS-R1a, was expressed at this site. Stimulation with ghrelin or GHS did not change basal or ACTH-stimulated corticosterone release, in accordance with earlier data demonstrating a lack of direct hormone-releasing effects of GHS on the adrenal gland (5,36).

2.4   The effect on the HPA axis in pathophysiological conditions

The effect of GHS and ghrelin has been studied in patients with various endocrine diseases. In subjects with isolated idiopathic GHD, ACTH and cortisol responses were identical to control normal subjects (37). Two studies investigated the effect of GHS on patients with two different types of GHRH receptor mutations. Both observed a normal ACTH and cortisol response to GHS stimuli (38,39), suggesting that, as opposed to the GH effect, functional GHRH receptors are not necessary for the effects of GHS on the HPA axis.

The effect of GHS in patients with Cushing's disease showed an unexpectedly large ACTH response, which exceeded the response seen in normal subjects. Indeed, the responses seen in patients with corticotroph microadenomas (with a smaller effect in macroadenomas) exceeded even the exaggerated ACTH responses to CRH (40,41). More recently, ghrelin has been shown to robustly stimulate ACTH and cortisol release in patients with Cushing's disease (Figure 6-6), in both micro- and macroadenomas (1). It has been suggested that a GHS test could be used for the differential diagnosis between ectopic and pituitary ACTH excess. However, some ectopic ACTH tumors contain high concentrations of the GHS-R (42), so it is likely that patients harbouring such tumors would show similar responsivity to GHS as patients with Cushing's disease.

The effect of GHRP-6 was studied in healthy subjects after metyrapone administration (inhibition of cortisol synthesis causing high ACTH levels) and in subjects with Addison's disease (43,44). In these two conditions the

GH response to GHRP-6 does not change. The ACTH response in patients with Addison's disease to GHS is more pronounced than in normal subjects (44).

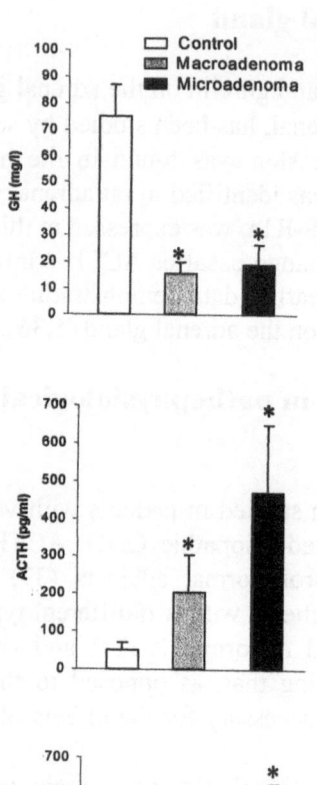

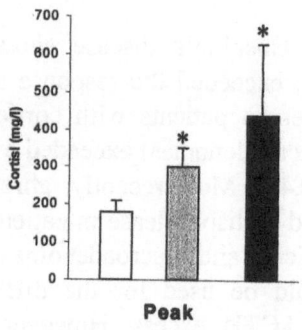

Figure 6-6. Mean±standard error peaks of ghrelin-induced GH, ACTH and cortisol secretion in control women and in Cushing's patients bearing pituitary micro- or macro-adenomas.
* p<0.05 vs.controls.
[redrawn from (1) with permission]

3.   EFFECT ON PROLACTIN RELEASE

GHS release PRL both in animal and human studies. A small PRL-releasing effect has been shown directly in pituitary cell cultures, so the major site of action may be either the pituitary or the hypothalamus (26). A rise in serum PRL has been reported in human studies with GHRP-6, GHRP-1, hexarelin and with the non-peptide compounds as well (Figure 6-1) (3,4,10,11,45). It is suggested that PRL release can be achieved by activation of the somato-mammotroph cells in the pituitary (10% of all the cells) (26,46). The PRL-releasing effect of GHRP-2 and hexarelin at both 1.0 and 2.0 µg/kg doses is significant, but peak PRL levels still remain within the normal range, and the effect is considerably smaller than the effect of thyrotropin releasing hormone (15). Sixteen weeks of subcutaneous administration of 1.5 µg/kg hexarelin to elderly subjects did not cause a significant change in PRL levels at the end of the treatment period (22). In a one year study using MK-0677, 3 patients were discontinued due to high PRL levels but the effect at one year on the rest of the treated patients has not been reported (24).

The effect of iv ghrelin 1.0 µg/kg (equivalent to 3×10^{-10} mol/kg) on peak PRL levels (14.9±2.2 µg/l) was significantly higher than that of the hexarelin 1.0 µg/kg (equivalent to 9×10^{-8} mol/kg) (10.9±0.9 µg/l) (Figure 6-2) (17).

3.1   The effect of ghrelin on PRL release in pathophysiological conditions

The effect of GHS has been studied in patients with pituitary adenomas. GHS stimulate GH release in patients with acromegaly, but do not stimulate further PRL release in patients with prolactinomas (47-51). However, *in vitro* secretion of PRL by GHS was demonstrated in human prolactinomas (52). The GH response in patients with microprolactinomas is blunted, according to one study (50), but was similar to normal subjects in another (51). Both GHRH- and GHRP-6-induced GH responses are attenuated in patients with macroprolactinomas, and synergism is present between the two agents. After bromocriptine treatment and tumor shrinkage, the GH responsiveness normalizes (51).

4.   EFFECT ON THE THYROID AXIS

GHS do not influence the level of thyroid stimulating hormone (TSH) according to a number of studies (4,13,45); however one group reported a drop of TSH levels after GHRP-1 (12) or hexarelin (53) administration. In a study involving healthy volunteers and the administration of placebo and hexarelin, a mild fall in TSH levels was observed on all study days (Figure 6-1), probably reflecting the well-known circadian rhythm of TSH (54). Neither the area under the curve, nor the nadir TSH levels, were significantly different between any of the treatment days. Similarly discrepant results have been reported after ghrelin administration, with either no change (55) or a small fall in TSH levels (56). Interestingly, intracerebroventricular ghrelin injection caused a decrease in TSH levels in one rat study (57). The influence of thyroid status was studied on ghrelin levels in rats (58). Circulating ghrelin and ghrelin mRNA expression in the stomach was increased in hypothyroidism while it was decreased in hyperthyroidism.

Ghrelin and GHS-R mRNA has been found to be expressed in the adult and fetal thyroid glands (35,59), although others failed to find expression in adult normal thyroid tissue (60). Binding sites for GHS were also identified in thyroid tissues (61). Medullary and follicular carcinoma tissue expressed ghrelin at a higher level than normal tissue (59,60), although this was not seen in all studies (62). An antiproliferative effect of GHS was observed in follicular carcinoma cell lines using [^{3}H]thymidine incorporation (62).

5.   EFFECT ON THE GONADOTROPH AXIS

Neither GHS nor ghrelin changed the levels of luteinizing hormone (LH), follicle stimulating hormone, estrogen or testosterone levels in acute or chronic human studies. A recent report on intracerebroventricular administration of ghrelin in gonadectomized and estrogen-replaced rats found an inhibitory effect on LH levels, primarily on pulse frequency, suggesting an action on the hypothalamus (63). As NPY has an inhibitory action on pulsatile LH secretion it is possible that this action occurs via the NPY stimulating effect of ghrelin.

Ghrelin binding sites were identified in the testis and ovaries (61), and ghrelin mRNA was detected in testis (35,64). Immunostaining detected ghrelin positivity in interstitial Leydig cells, while ghrelin treatment inhibited human chorionic gonadotrophin-induced testosterone secretion (64). Ghrelin expression in the testis was not influenced by GH deficiency or abnormal thyroid function but was inhibited by hypophysectomy and restored by human choriogonadotrophin injection suggesting dependence on LH activity (65). Ghrelin and GHS-R1a has been detected both at the mRNA and the protein level in prostate cancer cell lines where a proliferation stimulating effect has been observed (66).

Immunostaining detected ghrelin in the ovary especially in the interstitial cells and the mature corpus luteum (67). The GHS-R1a has also been identified in the corpus luteum as well as in the oocytes (67).

Ghrelin has also been identified in the human placenta, in cytotrophoblast cells, but its exact role is unclear (35,68). High ghrelin levels are detectable in the cord blood of newborn babies (69), suggesting that ghrelin might have effects on feeding behavior even at this stage. Some, but not all, studies suggest an influence of sex on ghrelin levels (70). Stomach ghrelin mRNA expression is also similar in male and female rats and is not influenced by either orchidectomy or ovariectomy (71).

Several recent reports investigated ghrelin levels in subjects with polycystic ovary syndrome (PCOS). Two studies suggest that ghrelin levels are lower in subjects with PCOS compared to body mass index (BMI)-matched controls (72,73). One of these studies especially noted discrepant ghrelin levels in subjects with insulin resistance (72), while in the other there was a negative correlation with androstendione levels (73). Whether low ghrelin in PCOS is a cause or the consequence of insulin resistance awaits further investigation, but is clearly of great interest as the disorder is extremely common and current therapeutic options are limited. A third study, however, found no difference in ghrelin levels between patients and age and BMI-matched controls (74).

6. GHRELIN AND GASTROINTESTINAL HORMONES

Both ghrelin and the GHS-R1a have been identified in the pancreas (35). Ghrelin and insulin interactions are discussed in detail elsewhere in this book. The effect of iv ghrelin injection on gastrin, SRIF and pancreatic polypeptide has been studied in humans (75): while there was no change in gastrin levels, both SRIF and pancreatic polypeptide levels increased considerably, further supporting the role of ghrelin in gut physiology.

Ghrelin and ghrelin receptors have been identified in neuroendocrine tumors including insulinomas, gastrinomas, glucagonomas and ACTH-secreting ectopic tumors (35,42,76-78). Ghrelin was found to be expressed in a wide variety of gut endocrine tumors (79,80). The pathophysiological relevance of these findings is currently unclear. Ghrelin has been found to be pro-proliferative in some systems while antiproliferative in others, so the effect of locally produced ghrelin needs to be investigated in individual tissue and tumor types.

7.   CONCLUSIONS

Ghrelin and GHS are potent stimulators of the HPA axis at a hypothalamic level, probably involving the stimulated release of both CRH and AVP. This effect, at least in part, may occur via the stimulation of hypothalamic NPY. In patients with ACTH-secreting tumors, the GHS-R may be expressed by the tumor, and thus such patients may show an unexpectedly enhanced response of ACTH and cortisol to ghrelin and GHS. It is currently unclear as to whether it is possible to completely dissociate the effect of GHS on the HPA axis and GH, but certainly the HPA effect is lost or attenuated on chronic treatment. A stimulatory effect on PRL may involve both pituitary and hypothalamic sites of action. Evidence for the involvement of the ghrelin system in the regulation of the thyroid and reproductive axes is much less robust, but local effects of ghrelin on these tissues are possible. Studies on placenta and in the polycystic ovarian system look highly interesting, and are clearly worthy of further study.

ACKNOWLEDGEMENTS

Márta Korbonits is supported by the Medical Research Council.

REFERENCES

1.   Leal-Cerro A, Torres E, Soto A, et al. Ghrelin is no longer able to stimulate growth hormone secretion in patients with Cushing's syndrome but instead induces exaggerated corticotropin and cortisol responses. Neuroendocrinol. 2002; 76(6):390-6.
2.   Bowers CY, Momany FA, Reynolds GA, Hong A. On the *in vitro* and *in vivo* activity of a new synthetic heptapeptide that acts on the pituitary to specifically release growth hormone. Endocrinology. 1984; 114:1537-45.
3.   Ghigo E, Arvat E, Gianotti L, et al. Growth hormone-releasing activity of Hexarelin, a new synthetic hexapeptide, after intravenous, subcutaneous, and oral administration in man. J Clin Endocrinol Metab. 1994; 78:693-8.
4.   Gertz BJ, Barrett JS, Eisenhandler R, et al. Growth hormone response in man to L-692,429, a novel nonpeptide mimic of growth hormone-releasing peptide-6. J Clin Endocrinol Metab. 1993; 77:1393-7.
5.   Thomas GB, Fairhall KM, Robinson ICAF. Activation of the hypothalamo-pituitary-adrenal axis by the growth hormone (GH) secretagogue, GH-releasing peptide-6, in rats. Endocrinology. 1997; 138(4):1585-91.
6.   Korbonits M, Grossman AB. Growth hormone-releasing peptide and its analogues; novel stimuli to growth hormone release. Trends Endocrinol Metab. 1995; 6:43-9.
7.   Cheng K, Chan WW, Barreto A, Jr., Convey EM, Smith RG. The synergistic effects of His-D-Trp-Ala-Trp-D-Phe-Lys-NH2 on growth hormone(GH)-releasing factor-

stimulated GH release and intracellular adenosin 3',5'-monophosphate accumulation in rat pituitary cell culture. Endocrinology. 1989; 124:2791-8.

8. Shimon I, Yan X, Melmed S. Human fetal pituitary expresses functional growth hormone-releasing peptide receptors. J Clin Endocrinol Metab. 1998; 83:174-8.

9. Jacks T, Hickey GJ, Taylor J, et al. Effects of acute and repeated intravenous administration of L-692,585, a novel nonpeptidyl growth hormone secretagogue, on plasma growth hormone, ACTH, cortisol, prolactin, Thyroxin (T4), insulin and IGF-1 levels in beagles. Program of the 76th Annual Meeting of the Endocrine Society, Anaheim, CA, 1994; p 365 (abstract).

10. Bowers CY, Reynolds GA, Durham D, Barrera CM, Pezzoli SS, Thorner MO. Growth hormone (GH)-releasing peptide stimulates GH release in normal men and acts synergistically with GH-releasing hormone. J Clin Endocrinol Metab. 1990; 70:975-82.

11. Hayashi S, Okimura Y, Yagi H, et al. Intranasal administration of His-D-Trp-Ala-Trp-D-Phe-LysNH2 (growth hormone releasing peptide) increased plasma growth hormone and insulin-like growth factor-I levels in normal men. Endocrinol Jpn. 1991; 38:15-21.

12. Laron Z, Bowers CY, Hirsch D, et al. Growth hormone-releasing activity of growth hormone-releasing peptide-1 (a synthetic heptapeptide) in children and adolescents. Acta Endocrinol (Copenh). 1993; 129:424-6.

13. Imbimbo BP, Mant T, Edwards M, et al. Growth hormone releasing activity of hexarelin in humans: a dose-response study. Eur J Clin Pharmacol. 1994; 46:421-5.

14. Bowers CY, Alster DK, Frentz JM. The growth hormone-releasing activity of a synthetic hexapeptide in normal men and short statured children after oral administration. J Clin Endocrinol Metab. 1992; 74:292-8.

15. Arvat E, Di Vito L, Maccagno B, et al. Effects of GHRP-2 and hexarelin, two synthetic GH-releasing peptides, on GH, prolactin, ACTH and cortisol levels in man. Comparison with the effects of GHRH, TRH and hCRH. Peptides. 1997; 18:885-91.

16. Clark RG, Thomas GB, Mortensen DL, et al. Growth hormone secretagogues stimulate the hypothalamic- pituitary-adrenal axis and are diabetogenic in the Zucker diabetic fatty rat. Endocrinology. 1997; 138(10):4316-23.

17. Arvat E, Maccario M, Di Vito L, et al. Endocrine activities of ghrelin, a natural GH secretagogue, in humans: comparison and interactions with hexarelin, a non natural peptidyl GHS, and GH-releasing hormone. J Clin Endocrinol Metab. 2001; 86:1169-74.

18. Korbonits M, Trainer PJ, Besser GM. The effect of an opiate antagonist on the hormonal changes induced by hexarelin. Clin Endocrinol (Oxf). 1995; 43:365-71.

19. Raun K, Sehested Hansen B, et al. Ipamorelin, the first selective growth hormone secretagogue. Eur J Endocrinol. 1998; 139:552-61.

20. Zdravkovic M, Sogaard B, Ynddal L, et al. The pharmacokinetics, pharmacodynamics, safety and tolerability of a single dose of NN703, a novel orally active growth hormone secretagogue in healthy male volunteers. Growth Horm IGF Res. 2000; 10:193-8.

21. Svensson J, Monson JP, Jorgensen JO, et al. Oral administration of the growth hormone (GH) secretagogue NN703 in adult patients with GH deficiency. Clin Endocrinol (Oxf). 2003; 58:572-80.

22. Rahim A, O'Neill PA, Shalet SM. The effect of chronic hexarelin administration on the pituitary-adrenal axis and prolactin. Clin Endocrinol. 1999; 50:77-84.

23. Chapman IM, Bach MA, Van Cauter E, et al. Stimulation of the growth hormone (GH)-insulin-like growth factor axis by daily oral administration of a GH-secretagogue (MK-0677) in healthy elderly subjects. J Clin Endocrinol Metab. 1996; 81:4249-57.

24. Murphy MG, Weiss S, McClung M, et al. Effect of alendronate and MK-677 (a growth hormone secretagogue), individually and in combination, on markers of bone turnover

and bone mineral density in postmenopausal osteoporotic women. J Clin Endocrinol Metab. 2001; 86(3):1116-25.

25. Massoud AF, Hindmarsh PC, Brook CGD. Interaction of the growth hormone releasing peptide hexarelin with somatostatin. Clin Endocrinol (Oxf). 1997; 47:537-47.

26. Cheng K, Chan WWS, Butler B, Wei L, Smith RG. A novel non-peptidyl growth hormone secretagogue. Horm Res. 1993; 40:109-15.

27. Korbonits M, Little JA, Forsling ML, et al. The effect of growth hormone secretagogues and neuropeptide Y on hypothalamic hormone release from acute rat hypothalamic explants. J Neuroendocrinol. 1999; 11:521-8.

28. Wren AM, Small CJ, Fribbens CV, et al. The hypothalamic mechanisms of the hypophysiotropic action of ghrelin. Neuroendocrinol. 2002; 76(5):316-24.

29. Mozid AM, Tringali G, Forsling ML, et al. Ghrelin is released from rat hypothalamic explants and stimulates corticotrophin-releasing hormone and arginine-vasopressin. Horm Metab Res. 2003; 35:455-9.

30. Korbonits M, Kaltsas G, Perry LA, et al. The growth hormone secretagogue hexarelin stimulates the hypothalamo-pituitary-adrenal axis via arginine vasopressin. J Clin Endocrinol Metab. 1999; 84:2489-95.

31. Arvat E, Maccagno B, Ramunni J, et al. Hexarelin, a synthetic growth hormone-releasing peptide, shows no interaction with corticotropin-releasing hormone and vasopressin on adrenocorticotropin and cortisol secretion in humans. Neuroendocrinol. 1997; 66:432-8.

32. Bennett PA, Thomas GB, Howard AD, et al. Hypothalamic growth hormone secretagogue-receptor (GHS-R) expression is regulated by growth hormone in the rat. Endocrinology. 1997; 138:4552-7.

33. Dickson SL, Leng G, Dyball REJ, Smith RG. Central actions of peptide and non-peptide growth hormone secretagogues in the rat. Neuroendocrinol. 1995; 61:36-43.

34. Ishizaki S, Murase T, Sugimura Y, et al. Role of ghrelin in the regulation of vasopressin release in conscious rats. Endocrinology. 2002; 143(5):1589-93.

35. Gnanapavan S, Kola B, Bustin SA, et al. The tissue distribution of the mRNA of ghrelin and subtypes of its receptor, GHS-R, in humans. J Clin Endocrinol Metab. 2002; 87(6):2988-91.

36. Barreiro ML, Pinilla L, Aguilar E, Tena-Sempere M. Expression and homologous regulation of GH secretagogue receptor mRNA in rat adrenal gland. Eur J Endocrinol. 2002; 147(5):677-88.

37. Aimaretti G, Baffoni C, Broglio F, et al. Endocrine responses to ghrelin in adult patients with isolated childhood-onset growth hormone deficiency. Clin Endocrinol (Oxf). 2002; 56(6):765-71.

38. Maheshwari HG, Rahim A, Shalet SM, Baumann G. Selective lack of growth hormone (GH) response to the GH-releasing peptide hexarelin in patients with GH-releasing hormone receptor deficiency. J Clin Endocrinol Metab. 1999; 84:956-9.

39. Gondo RG, Aguiar-Oliveira MH, Hayashida CY, et al. Growth hormone-releasing peptide-2 stimulates GH secretion in GH-deficient patients with mutated GH-releasing hormone receptor. J Clin Endocrinol Metab. 2001; 86(7):3279-83.

40. Arvat E, Giordano R, Ramunni J, et al. Adrenocorticotropin and cortisol hyperresponsiveness to hexarelin in patients with Cushing's disease bearing a pituitary microadenoma, but not in those with macroadenoma. J Clin Endocrinol Metab. 1998; 83:4207-11.

41. Ghigo E, Arvat E, Ramunni J, et al. Adrenocorticotropin- and cortisol-releasing effect of hexarelin, a synthetic growth hormone-releasing peptide, in normal subjects and patients with Cushing's syndrome. J Clin Endocrinol Metab. 1997; 82(8):2439-44.

42. Korbonits M, Jacobs RA, Aylwin SJB, et al. Expression of the growth hormone secretagogue receptor in pituitary adenomas and other neuroendocrine tumors. J Clin Endocrinol Metab. 1998; 83:3624-30.

43. Pinto ACAR, Silva MRD, Martins MR, Brunner E, Lengyel AMJ. Effects of short-term glucocorticoid deprivation on growth hormone (GH) response to GH-releasing peptide-6: Studies in normal men and in patients with adrenal insufficiency. J Clin Endocrinol Metab. 2000; 85:1540-4.

44. Martins MR, Pinto ACAR, Brunner E, Silva MRD, Lengyel AMJ. GH-releasing peptide (GHRP-6)-induced ACTH release in patients with Addison's disease: Effect of glucocorticoid withdrawal. J Endocrinol Invest. 2003; 26:143-7.

45. Ilson BE, Jorkasky DK, Curnow RT, Stote RM. Effect of a new synthetic hexapeptide to selectively stimulate growth hormone release in healthy human subjects. J Clin Endocrinol Metab. 1989; 69:212-4.

46. Smith RG, Van der Ploeg LHT, Howard AD, et al. Peptidomimetic regulation of growth hormone secretion. Endocrine Rev. 1997; 18:621-45.

47. Alster DK, Bowers CY, Jaffe CA, Ho PJ, Barkan AL. The growth hormone (GH) response to GH-releasing peptide (His- DTrp-Ala-Trp-DPhe-Lys-NH2), GH-releasing hormone, and thyrotropin- releasing hormone in acromegaly. J Clin Endocrinol Metab. 1993; 77:842-5.

48. Hanew K, Utsumi A, Sugawara A, Shimizu Y, Abe K. Enhanced GH responses to combined administration of GHRP and GHRH in patients with acromegaly. J Clin Endocrinol Metab. 1994; 78:509-12.

49. Popovic V, Damjanovic S, Micic D, Petakov M, Dieguez C, Casanueva FF. Growth hormone (GH) secretion in active acromegaly after the combined administration of GH-releasing hormone and GH-releasing peptide-6. J Clin Endocrinol Metab. 1994; 79:456-60.

50. Ciccarelli E, Grottoli S, Razzore P, et al. Hexarelin, a growth hormone releasing peptide, stimulates prolactin release in acromegalic but not in hyperprolactinaemic patients. Clin Endocrinol (Oxf). 1996; 44:67-71.

51. Popovic V, Simic M, Ilic L, et al. Growth hormone secretion elicited by GHRH, GHRP-6 or GHRH plus GHRP-6 in patients with microprolactinoma and macroprolactinoma before and after bromocriptine therapy. Clin Endocrinol (Oxf). 1998; 48:103-8.

52. Adams EF, Huang B, Buchfelder M, et al. Presence of growth hormone secretagogue receptor messenger ribonucleic acid in human pituitary tumors and rat GH3 cells. J Clin Endocrinol Metab. 1998; 83:638-42.

53. Laron Z, Frenkel J, Gil-Ad I, et al. Growth hormone releasing activity by intranasal administration of a synthetic hexapeptide (hexarelin). Clin Endocrinol (Oxf). 1994; 41:539-41.

54. Chan V, Jones A, Liendo-Ch P, McNielly A, Landon J, Besser GM. The relationship between circadian variations in circulating thyrotropin, thyroid hormones, and prolactin. Clin Endocrinol (Oxf). 1978; 9:337-49.

55. Takaya K, Ariyasu H, Kanamoto N, et al. Ghrelin strongly stimulates growth hormone release in humans. J Clin Endocrinol Metab. 2000; 85:4908-11.

56. Hataya Y, Akamizu T, Takaya K, et al. A low dose of ghrelin stimulates growth hormone (GH) release synergistically with GH-releasing hormone in humans. J Clin Endocrinol Metab. 2001; 86(9):4552-5.

57. Wren AM, Small CJ, Ward HL, et al. The novel hypothalamic peptide ghrelin stimulates food intake and growth hormone secretion. Endocrinology. 2000; 141(11):4325-8.

58. Caminos JE, Seoane LM, Tovar SA, Casanueva FF, Dieguez C. Influence of thyroid status and growth hormone deficiency on ghrelin. Eur J Endocrinol. 2002; 147(1):159-63.

59. Kanamoto N, Akamizu T, Hosoda H, et al. Substantial production of ghrelin by a human medullary thyroid carcinoma cell line. J Clin Endocrinol Metab. 2001; 86(10):4984-90.

60. Volante M, Allla E, Fulcheri E, et al. Ghrelin in fetal thyroid and follicular tumors and cell lines: expression and effects on tumor growth. Am J Pathol. 2003; 162(2):645-54.

61. Papotti M, Ghe C, Cassoni P, et al. Growth hormone secretagogue binding sites in peripheral human tissues. J Clin Endocrinol Metab. 2000; 85:3803-7.

62. Cassoni P, Papotti M, Catapano F, et al. Specific binding sites for synthetic growth hormone secretagogues in non-tumoral and neoplastic human thyroid tissue. J Endocrinol. 2000; 165(1):139-46.

63. Furuta M, Funabashi T, Kimura F. Intracerebroventricular administration of ghrelin rapidly suppresses pulsatile luteinizing hormone secretion in ovariectomized rats. Biochem Biophys Res Commun. 2001; 288(4):780-5.

64. Tena-Sempere M, Barreiro ML, Gonzalez LC, et al. Novel expression and functional role of ghrelin in rat testis. Endocrinology. 2002; 143(2):717-25.

65. Barreiro ML, Gaytan F, Caminos JE, et al. Cellular location and hormonal regulation of ghrelin expression in rat testis. Biol Reprod. 2002; 67(6):1768-76.

66. Jeffery PL, Herington AC, Chopin LK. Expression and action of the growth hormone releasing peptide ghrelin and its receptor in prostate cancer cell lines. J Endocrinol. 2002; 172(3):R7-R11.

67. Gaytan F, Barreiro ML, Chopin LK, et al. Immunolocalization of ghrelin and its functional receptor, the type 1a growth hormone secretagogue receptor, in the cyclic human ovary. J Clin Endocrinol Metab. 2003; 88(2):879-87.

68. Gualillo O, Caminos JE, Blanco M, et al. Ghrelin, a novel placental-derived hormone. Endocrinology. 2001; 142(2):788-94.

69. Chanoine JP, Yeung LP, Wong AC, Birmingham CL. Immunoreactive ghrelin in human cord blood: relation to anthropometry, leptin, and growth hormone. J Pediatr Gastroenterol Nutr. 2002; 35(3):282-6.

70. Barkan AL, Dimaraki EV, Jessup SK, Symons KV, Ermolenko M, Jaffe CA. Ghrelin secretion in humans is sexually dimorphic, suppressed by somatostatin, and not affected by the ambient growth hormone levels. J Clin Endocrinol Metab. 2003; 88:2180-4.

71. Schofl C, Horn R, Schill T, Schlosser HW, Muller MJ, Brabant G. Circulating ghrelin levels in patients with polycystic ovary syndrome. J Clin Endocrinol Metab. 2002; 87(10):4607-10.

72. Pagotto U, Gambineri A, Vicennati V, Heiman ML, Tschöp M, Pasquali R. Plasma ghrelin, obesity, and the polycystic ovary syndrome: correlation with insulin resistance and androgen levels. J Clin Endocrinol Metab. 2002; 87(12):5625-9.

73. Orio F, Jr., Lucidi P, Palomba S, et al. Circulating ghrelin concentrations in the polycystic ovary syndrome. J Clin Endocrinol Metab. 2003; 88(2):942-5.

74. Moller N, Nygren J, Hansen TK, Orskov H, Frystyk J, Nair KS. Splanchnic release of ghrelin in humans. J Clin Endocrinol Metab. 2003; 88(2):850-2.

75. Korbonits M, Bustin SA, Kojima M, et al. The expression of the growth hormone secretagogue receptor ligand ghrelin in normal and abnormal human pituitary and other neuroendocrine tumors. J Clin Endocrinol Metab. 2001; 86(2):881-7.

76. Volante M, Allla E, Gugliotta P, et al. Expression of ghrelin and of the GH secretagogue receptor by pancreatic islet cells and related endocrine tumors. J Clin Endocrinol Metab. 2002; 87(3):1300-8.
77. Iwakura H, Hosoda K, Doi R, et al. Ghrelin expression in islet cell tumors: augmented expression of ghrelin in a case of glucagonoma with multiple endocrine neoplasm type I. J Clin Endocrinol Metab. 2002; 87(11):4885-8.
78. Papotti M, Cassoni P, Volante M, Deghenghi R, Muccioli G, Ghigo E. Ghrelin-producing endocrine tumors of the stomach and intestine. J Clin Endocrinol Metab. 2001; 86(10):5052-9.
79. Rindi G, Savio A, Torsello A, et al. Ghrelin expression in gut endocrine growths. Histochem Cell Biol. 2002; 117(6):521-5.

Chapter 7

GHRELIN FOOD INTAKE AND ENERGY BALANCE

Matthias H. Tschöp, Tamara Castañeda & Uberto Pagotto[1]
Department of Pharmacology, German Institute of Human Nutrition, Potsdam-Rehbrücke, Germany; [1]Department of Endocrinology, S. Orsola-Malpighi Hospital, Bologna, Italy

Abstract: In order to develop an effective pharmacological treatment for obesity, an endogenous factor that promotes a positive energy balance by increasing appetite and decreasing fat oxidation could represent the drug target scientists have been looking for. The recently discovered gastric endocrine agent ghrelin, which appears to be the only potent hunger-inducing factor to naturally circulate in our blood stream, was discovered in 1999. Since then the acylated peptide hormone ghrelin has evolved from an endogenous growth hormone secretagogue to a regulator of energy balance to a pleiotropic hormone with multiple sources, numerous target tissues and most likely several physiological functions. Although neither the exact mechanism of action by which ghrelin increases food intake and adiposity is known, nor the putatively differential effects of brain-derived and stomach-derived ghrelin on energy homeostasis have been determined, blocking or neutralizing ghrelin action still seems one of the more reasonable pharmacological approaches to reverse a chronically positive energy balance. However, based on growing experience with compounds targeting the neuroendocrine regulation of energy balance, it is quite possible that a ghrelin antagonist will either fail to cure obesity due to the existence of compensatory mechanisms or undesired effects might reveal the true biological function of ghrelin (e.g. cardiovascular mechanisms, anti-proliferative effects, reproduction).

Key words: food intake, energy balance

The discovery that a novel receptor and its endogenous ligand are controlling several essential physiological functions naturally sparked the interest of the scientific community. Thus, the discovery of the novel hormone ghrelin in 1999 (1), much like leptin in 1994 (2), triggered an enormous amount of research studies focusing on this new endocrine player.

The discovery of ghrelin has often been referred to as a classical example of *reverse pharmacology,* since synthetic ghrelin analogues (growth hormone secretagogues, GHS) were described first, while the natural receptor (GHS receptor type 1a (GHS-R1a)), its endogenous ligand (ghrelin) and the physiological role of ghrelin in energy balance regulation have been discovered stepwise over the last 25 years (1,3,4). The pharmacological applicability of ghrelin's endogenous counterpart leptin has been somewhat dissapointing, at least with respect to the treatment of diet-induced human obesity (5). The ghrelin pathway could now offer new therapeutic perspectives as an anti-obesity target, since ghrelin effectively induces hunger in humans and is the only known peripheral orexigenic signal counterbalancing a variety of pancreatic, gastrointestinal and adiypocite-derived satiety factors via its appetite-stimulating effects (6). This review attempts to comprehensively evaluate the potential of ghrelin and its receptor as a drug target, focusing on its role in appetite regulation and energy homeostasis, but also shortly discussing a potential role for ghrelin, ghrelin mimetics or ghrelin antagonists in specific disease entities such as the Prader-Willi Syndrome (PWS).

The fatty acid (n-octanoyl) side chain at serine 3, a biochemical feature which is essential for ghrelin's bioactivity, signifies this gastrointestinal peptide hormone as an endogenous factor unique in mammalian biology (7). While cleaving more than 50% of its 28 amino residues starting from the C-terminal end of the ghrelin molecule (down to less than 14 amino residues) hardly influences binding or activation of its receptor *in vitro*, minor modifications of the postranslationally added n-octanoic acid already impair receptor binding or activation (8). Amino acid 28 (arginine) is naturally cleaved up to an until now unknown percentage of stomach-derived circulating ghrelin molecules, resulting in a 27 amino acids long peptide, which is still bioactive. As reported very recently, the octanoyl side chain occurs in at least two different sizes (C8 and C10), while the identity of a putative acyl-transferase that is presumably located in the stomach and should be responsible for the "activation" of ghrelin via octanoylation, is still unknown (9). Degradation processes are also believed to mainly involve enzymatic processes, since the half-life of ghrelin in circulation is estimated to be between 5 and 15 minutes. The only factor that has actually been shown to "deactivate" ghrelin is high density lipoprotein, which can bind and des-acylate ghrelin to a significant extent, thereby depriving it of its ability to bind and activate the ghrelin receptor GHS-R1a (10).

While an enormous amount of data on ghrelin biology and physiology is currently emerging, it should not be overlooked that additional ghrelin receptors (in addition to the GHS-R1a and 1b), as well as other endogenous ligands of the GHS-R1a, might exist and could play a relevant role (11). The vast majority of currently available data are either focusing on the GHS-R1a, or are extrapolating from changes of the overall concentration of circulating (bio-active and bio-inactive) ghrelin peptide, that for now is the best available, easily accessible, surrogate parameter for the activity level of the ghrelin pathway. While the detection of relative differences in total circulating ghrelin levels (predominantly representing bio-inactive peptide) between disease states or in response to physiological challenges can be regarded as useful, one should be careful not to over-interpret these data. Once more sophisticated methods to monitor plasma concentrations of active ghrelin as well as expression and activation levels of specific ghrelin receptor subtypes are available, substantial parts of the current view on ghrelin (patho-) physiology might have to be adjusted, if not corrected. The extent and magnitude of ghrelin action must involve multiple regulatory levels, which might, at least in part, be independent from each other. Relevant mechanisms include: a) the regulation of transcription and translation of the ghrelin gene; b) the level of enzymatic activity of the putative acyl transferase that is responsible for the post-translational octanoylation of the ghrelin molecule; c) secretion rates of the bioactive ghrelin molecule; d) putative enzymatic processes de-activating circulating ghrelin; e) possible influence of ghrelin binding proteins on the hormones' bioactivity (e.g. binding of high density lipoproteins); f) variable accessibility of target tissue (i.e. blood-brain barrier transport); g) clearance or degradation of ghrelin by kidney or liver passage; h) circulating concentration of additional endogenous ligands or other possibly cross-reacting hormones; i) the amount of expression of ghrelin receptor(s) in target tissue and j) their sensitivity at the level of intracellular signaling mechanisms.

Ghrelin is primarily expressed in the stomach and the upper intestinal tract (12). However, recent studies using polymerase chain reaction (PCR) amplification techniques have revealed the potential localization of ghrelin mRNA in several other tissues, such as the kidneys, immuno-competent blood cells, placenta, testicles, ovaries, pancreas, pituitary, hypothalamus and other tissues (13). In respect to ghrelin's putative role in the regulation of energy homeostasis, a localization in the hypothalamus seems to be particularly interesting. Using antibodies as well reverse transcriptase-PCR (RT-PCR), a uniquely distributed hypothalamic group of mostly bipolar neurons has recently been identified as producing small amounts of ghrelin (14). These neurons are not co-localized with any known centrally expressed hormone or neuropeptide, but they intriguingly do project directly to several previously identified hypothalamic appetite control centers. Ghrelin receptor expression and binding can furthermore be localized in multiple hypothalamic areas in direct neighborhood to neuropeptide Y (NPY), agouti-related protein (AgRP), pro-opiomelanocortin (POMC), gamma-aminobutyric acid and other neuropeptides and neurotransmitters substantially involved in appetite control. Ghrelin expression can be found in neurons situated closely to, but not connected with, the previously mentioned ones. These neuroanatomical findings, complemented by electrophysiology studies, provide evidence for the existence of a central circuit regulating appetite that involves ghrelin as a key-modulator (14).

The main source of circulating endogenous ghrelin is still the stomach (15). Several possibilities regarding the role of central versus peripheral ghrelin in the regulation of food intake are thinkable and have to be proven or disproven in the future, possibly by using sophisticated approaches such as mice after gastrectomy or with tissue-specific disruption of the ghrelin gene:

1. gastroenterally-derived circulating ghrelin could be co-modulating central networks regulating energy balance together with hypothalamically-derived ghrelin after crossing the blood-brain barrier or via neural projection from areas that are not protected by the blood-brain barrier (circumventricular organs, e.g., median-eminence);

2. gastroenterally-derived circulating ghrelin could be responsible for peripheral effects of ghrelin including direct effects on endocrine axes at the pituitary level, cardiovascular, anti-proliferative or adipocyte specific effects, while centrally derived ghrelin mainly modulates energy balance control circuits;

3. gastroenterally-derived ghrelin co-modulates the central regulation of energy balance at the gastric level via the vagal nervous system and the brainstem.

There appears to be no reason why ghrelin could not act in parallel via paracrine mechanisms in the brain, endocrine mechanisms via circumventricular organs or after crossing the blood-brain barrier and via the parasympathetic nervous system at the gastrointestinal level to regulate energy balance (16).

Ghrelin administration induces adiposity (17,18), raises the respiratory quotient (reflecting reduced fat utilization) (19) and suppresses spontaneous locomotor activity in rodents (Tang-Christensen & Tschöp, unpublished results). Neutralization studies with polyclonal ghrelin antibodies yielded encouraging results, showing decreased food intake in rodents after intracerebroventricular injection (20). These data were confirmed by a transgenic rat model overexpressing antisense-oligonucleotides against the ghrelin receptor GHS-R1a, where decreased food intake and lower body fat were observed as a consequence (21). There is clear evidence that, in spite of a relatively short half life, ghrelin administration in physiologically relevant doses is inducing a positive energy balance (22). Ghrelin affects body weight and food intake more than a thousand-fold more potently following central administration, strongly supporting the hypothesis that ghrelin influences energy homeostasis predominantly via the modulation of central mechanisms (23). Wherever therefore the decisive endogenous amount of ghrelin that regulates energy balance is mainly derived from, blocking its endogenous actions acutely as well as chronically will at least teach a valuable physiology lesson, if not pave the way for the development of a new drug.

Mainly secreted by gastric A/X-like cells within the oxyntic glands, the half life of ghrelin is relatively short (5-15 minutes) and less than 20% of the circulating immunoreactive ghrelin appears to be octanoylated and therefore bioactive (24,25). Gastrointestinal X/A-like cells represent about a quarter of all endocrine cells in the oxyntic mucosa, while other cells within these glands, such as histamine-rich enterochromaffin-like cells (ca. 70%) and D-(somatostatin) cells (10%), are ghrelin-negative (12). From the stomach to the colon ghrelin is found with caudally decreasing expression (26). Ghrelin containing entero-endocrine cells mostly have no continuity with the lumen, probably respond to physical and/or chemical stimuli from the baso-lateral side, and are closely associated with the capillary network running through the lamina propria. Ghrelin secreting cells occur as open- and close-type cells (open or closed towards the stomach lumen) with the number of open type cells gradually increasing in the direction from the stomach to the lower gastrointestinal tract (12). A closer look at the structural and functional relationship between ghrelin and its receptor and the receptor and structure of motilin and its receptor suggests that a larger family of peptide hormones is co-modulating gastrointestinal motility, appetite, secretion of pituitary hormones and other physiological processes (27). This same peripheral endocrine network most likely also includes gastrointestinal hormones such as cholecystokinine, peptide YY_{1-36} and peptideYY_{3-36}, glucagon-like peptide 1, gastric inhibitory peptide and others (28). Both ghrelin peptide secretion and ghrelin mRNA expression are regulated according to metabolic challenges (15). Acute as well as chronic periods of food deprivation going from mild to severe (e.g. fasting) increase ghrelin peptide levels as well as ghrelin mRNA concentration, whereas re-feeding reduces ghrelin peptide secretion as well as ghrelin mRNA expression (29).

While a classical endocrine role for ghrelin as a peptide hormone that is secreted into a capillary network is evident, local, paracrine activities of ghrelin might play an additional role. Removal of the stomach or the acid-producing part of the stomach in rodents reduces serum ghrelin concentration by ca. 80%, further supporting the notion that the stomach is the main source of this endogenous GHS-R ligand (15). That total plasma ghrelin is hardly detectable following gastric bypass surgery, Cummings and coworkers interpreted as a "shut down" of gastric ghrelin secretion due to complete lack of contact with ingested nutrients (30).

Ghrelin mRNA and ghrelin peptide have also been detected in rat and human placenta (31). Here, ghrelin is expressed predominantly in cytotrophoblast cells and very sporadically in syncytiotrophoblast cells. A pregnancy related time-course, represented by an early rise of ghrelin expression in the third week and decreasing stages in the latest stages of gestation, as well as still detectable presence of ghrelin at term, was found in rats. In human placenta, ghrelin is mainly expressed in the first half of pregnancy and is not detectable at term, while a putative involvement of ghrelin in fetal-maternal interaction via autocrine, paracrine, or endocrine mechanisms still remains to be shown. Small concentrations of ghrelin are found in the pancreas, where ghrelin immunoreactivity has been localized in a sub-group of endocrine cells that are also immuno-positive for pancreostatin. Ongoing, partially contradictory studies suggest that ghrelin positive cells must belong to either the pancreatic A-cells (16), pancreatic B-cells (32), or pancreatic non-A-non-B-cells (33).

In normal pituitary cells as well as in pituitary tumors ghrelin mRNA expression and ghrelin immunopositive cells were detected in addition to the known presence of GHS-R in pituitary cells. This suggests a possible autocrine or paracrine role for hypophyseal ghrelin, although only ca. 5% of the detected ghrelin peptide derived from the pituitary has been found to be octanoylated (34). Using RT-PCR methodology, small amounts of ghrelin were detected in the adrenal glands, esophagus, adipocytes, gall bladder, muscle, myocardium, ovary, prostate, skin, spleen, thyroid, blood vessels, and liver (35). Prepro-ghrelin production was shown in rat mesangial cells and mouse podocytes, indicating the production of ghrelin in kidney, glomerulus and renal cells and suggesting possible paracrine roles for ghrelin in the kidney (36). Human ghrelin as well as GHS-R mRNA-expression was shown by RT-PCR and confirmed by DNA-sequencing in human T-lymphocytes, B-lymphocytes and neutrophils from venous blood of healthy volunteers. Cell type and maturity of the cells did not seem to have an influence on ghrelin production in immune cells (37). Interestingly, it has recently been shown that small molecule GHS have a considerable immune-enhancing effect (38). In summary, secreted ghrelin is expressed in its majority by the stomach, followed by still substantial concentrations deriving from lower parts of the gastrointestinal tract (39). While its physiological significance as a paracrine factor in extra-gastrointestinal tissue is the subject of ongoing studies, a classical endocrine role for extra-gastrointestinal ghrelin appears to be unlikely since ghrelin expression levels in other organs are relatively low in comparison. Published studies on the regulation of ghrelin expression were therefore primarily focused on gastric ghrelin. As a cautionary note it should be added however that studies on ghrelin expression or secretion in rodents are however not necessarily relevant for the physiological regulation of ghrelin in humans.

A very intriguing series of clinical studies indicates that each daily meal is followed by decreases of circulating ghrelin levels, most likely reflecting acutely reduced ghrelin secretion from the gastrointestinal tract (40,41). The authors speculate in addition that an observed pre-meal rise of circulating human ghrelin levels might reveal a role for ghrelin in meal initiation, which fits well with the observation that ghrelin administration in healthy volunteers causes hunger sensations (40). Ghrelin might also reflect the acute state of energy balance, signaling to the central nervous system (CNS) in times of food deprivation that increased energy intake and an energy-preserving metabolic state are desirable (22).

Only few determinants of circulating ghrelin concentration have been identified to date, including insulin, glucose, somatostatin and possibly growth hormone, leptin, melatonin and the parasympathetic nervous system tone (38). In several species (e.g. rodents, cows and humans) ghrelin mRNA-expression levels or circulating ghrelin levels have been shown to be increased by food deprivation and to be decreased postprandially (17,40,42). This phenomenon further supports the concept of ghrelin as an endogenous regulator of energy homeostasis that has apparently been preserved throughout species during evolution (13). Rat ghrelin expression can also be stimulated by insulin-induced hypoglycemia, leptin administration (29) and central leptin gene therapy (43). Ingestion of sugar suppresses ghrelin secretion in rats *in vivo*, indicating a possible direct inhibitory effect of glucose/caloric intake on ghrelin containing X/A-like cells in the oxyntic mucosa of the rat stomach but not excluding an additional insulin-mediated effect (17). That insulin is an independent determinant of the circulating ghrelin concentration has recently been shown by several study groups using hyperinsulinemic euglycemic studies in humans (44-48). These findings add further evidence for a connecting role of ghrelin between mechanisms governing energy balance and the regulation of glucose homeostasis. It remains however to be shown if postprandially occurring insulin peaks are sufficient to decrease circulating ghrelin levels, since hyperinsulinemic-euglycemic clamp studies causing decreased ghrelin secretion involve either supraphysiological or markedly prolonged (e.g. >120 min) periods of hyperinsulinemia (49). Further insight into the complex mechanisms regulating ghrelin secretion is based on studies showing an increase of circulating ghrelin levels in rats following surgical interventions such as vagotomy and hypophysectomy. Human GH-deficiency, however, does not seem to exhibit increased plasma ghrelin levels (50). On the other hand, administration of synthetic GH in rats decreases circulating ghrelin levels (51) and patients with acromegaly show low endogenous ghrelin levels (52). Contradictory observations could be caused by species-specific differences between rodents and humans. Another explanation would be that an acute, but not a chronic, change of GH levels modulates ghrelin concentration. A more exotic pathophysiological mechanism responsible for high circulating ghrelin levels is the production of ghrelin by endocrine tumors of the stomach and the intestine such as carcinoids (53).

In summary, ghrelin expression as well as ghrelin secretion are predominantly influenced by changes in energy balance and glucose homeostasis, followed by alterations of endocrine axes (e.g. increasing GH concentrations). Based on the currently available data, ghrelin therefore mainly seems to represent a molecular regulatory interface between energy homeostasis, glucose metabolism and physiological processes regulated by the classical endocrine axes such as growth and reproduction. One particular biological purpose of these multiple roles of ghrelin might be to ensure the provision of calories that GH requires for growth and repair.

Differing from earlier models expecting the endogenous ligand of the growth-hormone secretagogue receptor to govern growth hormone secretion (1), ghrelin is currently believed to have its main physiological role in the regulation of energy balance (22). As the only potent circulating orexigenic agent known to date, ghrelin triggers appetite and nutrient intake (17,54). Ghrelin might even represent the first "meal initiation factor" known (40). However, conclusive evidence that meal-related circadian changes of plasma ghrelin concentrations are responsible for the initiation of nutrient intake rather than representing an epi-phenomenon of trained meal patterns is still missing. Based on clinical investigations of meal related changes of plasma ghrelin levels and data generated by insulin- and glucose-clamp studies, plasma insulin as well as blood glucose levels are very likely to be involved in the general regulation of ghrelin secretion (46). Although hyperinsulinemic-euglycemic clamps have been repeatedly shown to decrease circulating ghrelin, it remains unclear if experimental conditions during clamp studies are comparable with the lower maximum peaks and the shorter duration of postprandial insulin levels (49). It can not be excluded that additional blood derived factors are responsible for meal-related changes in ghrelin concentrations or that gastrointestinal nutrient sensors are modulating ghrelin expression and secretion rates.

Although counterintuitive, the finding that circulating ghrelin concentrations are low in obesity not only mirrors earlier observations of hyperleptinemia in obesity, but may be explained by compensatory mechanisms aiming to communicate to central regulatory centers that energy stores are sufficiently filled (55). While it is not clear which signal communicates increased adipocyte size to ghrelin secreting cells (leptin, IL-6, adiponectin would be candidates), it has to be carefully investigated if other phenomena and symptoms that are frequently occuring in obesity (such as a frequently filled stomach or insulin resistance) contribute to hypoghrelinemia. Ghrelin gene polymorphisms have been described by several groups, linkage analysis studies however failed to prove a solid association between ghrelin and obesity (56-58). While diet-induced human obesity, as well as polygenic (e.g. Pima Indians) or monogenic (e.g. MC4-R defect) causes of human obesity all present with low plasma ghrelin levels (59), there is one group of severely obese patients where markedly increased plasma ghrelin levels have been observed. In patients with Prader-Willi syndrome, an impressive hunger syndrome along with morbid obesity and numerous other symptoms is caused by a missing part of the short arm of chromosome 15 and accompanied by 3-5-fold higher circulating ghrelin levels when compared to healthy controls (60). While the overlap between symptoms of PWS and effects of ghrelin administration is impressive, only treatment with a potent, but safe ghrelin antagonist compound will show if ghrelin is part of the pathogenesis in PWS (61). There is an ongoing discussion if increased plasma ghrelin levels are only a consequence of severe caloric restrictions which are a central part of treatment strategies for patients with PWS in an attempt to control their energy balance. The only other population where comparably high ghrelin levels have been reported are patients with cachexia or anorexia nervosa, where high ghrelin levels are believed to reflect a physiological compensation effort in response to either a chronically empty stomach or a markedly decreased fat mass (62). While circulating ghrelin levels are significantly lower in obese individuals, their ghrelin levels are still very substantial, when compared to nearly undetectable ghrelin concentrations in patients after gastric bypass surgery (30). The superior effectivity of this bariatric procedure is discussed as partially due to a *"knock-down"* of endogenous ghrelin secretion caused by the lack of stimulation of gastric cells by incoming nutrients. On the other hand there is a chance that an increase of endogenous ghrelin in response to diet-induced weight loss contributes to the high probability of obesity recidivism. Carefully conducted clinical studies are mandatory to solve this important question.

Ghrelin administration in rodents causes weight gain (17). This effect would not be as surprising if it was merely reflected by longitudinal growth or at least by an increase in lean mass, effects that one would expect to occur after stimulation of GH-secretion. However, a still growing body of data generated in rodents clearly shows that ghrelin-induced weight gain is based on accretion of fat mass without changes in longitudinal skeletal growth and with a tendency toward a decrease, rather than an increase of lean (muscle) mass (63). These findings have not only been confirmed by several groups but have also been repeated using synthetic ghrelin receptor (GHS receptor) agonists NNC 26-0161 (ipamorelin) (18), or GHRP-2 (19) and GHRP-6 (18). Changes in body weight induced by ghrelin administration become significant in rodents after no more than 48 h and are clearly visible at the end of two weeks. Changes in fat mass induced by GHS have been quantified using Dual Energy X-ray Absorptiometry measurements specifically adapted for analysis of rodent body composition and have also been confirmed using chemical carcass analysis (51), or by measuring the weight of omental and retroperitoneal fat pads. Energy balance is achieved when energy intake is equal to energy expenditure. A positive energy balance, leading to weight gain, occurs when calories ingested, digested and re-absorbed exceeds calories expended. Like leptin, but in an opposite manner, ghrelin administered in rodents influences both energy intake and metabolism to increase body fat and body weight (11).

Ghrelin's orexigenic action is comparable to that of NPY when administered centrally and is more effective than any other orexigenic agent (64). While peripherally injected GHS or ghrelin does have less impressive, predominantly acute and transient orexigenic effects, continuously administered ghrelin (3rd ventricle) causes potent and constant stimulation of appetite in rats (17). However, further studies (i.e., involving mice with tissue-specific disruption of the ghrelin gene) will have to prove the existence of an endogenous ghrelin-tone that supports ghrelin's putative relevance for physiological appetite regulation and metabolic control. Synthetic GHS (ghrelin receptor agonists) that have been shown to have orexigenic activity so far include GHRP-2, ipamorelin, GHRP-6, hexarelin, EP 50885, EP 40904 and EP92632 (65). Apart from an increase of food intake, other mechanisms can contribute to an increase in fat mass, such as a decrease in energy expenditure or reduced cellular fat oxidation. No significant changes of 24-hour total energy expenditure have been observed in rodents after ghrelin administration (17). Improved methodology to quantify and monitor energy expenditure in small animals is needed to clarify discrepancies in energy balance characterization, since lack of sufficient sensitivity could be one reason why small (but significant when chronically occurring) changes in energy expenditure are currently overlooked. One effect that was detected by indirect calorimetry is an increase of the respiratory quotient after ghrelin administration in rodents (19). It remains to be shown that this increase is independent from an increase in food intake. A raised respiratory quotient is interpreted as a shift from fat utilization to carbohydrate oxidation and has also been referred to as indicating changes in "nutrient partitioning" (66).

The network of neurons, neuropeptides, neurotransmitters and receptors controlling energy balance is an extremely complex, multi-centered system (67). Based on early surgical and chemical deletion studies in rodents, the hypothalamus has long been recognized as a crucial interface between afferent peripheral signals, other regulatory centers in the CNS and efferent pathways regulating energy balance in concert (68). In addition, the brainstem and other brain areas are emerging as equally important regulatory centers of energy homeostasis control (69). Ghrelin binds predominantly at NPY- and AgRP-coexpressing neurons in the arcuate nucleus, which are also known to express the ghrelin receptor GHS-R1a (14,70). Ghrelin has an increasing effect on both orexigenic neuropeptides, along with an indirect suppressing effect on POMC and cocaine-amphetamine regulated transcript expressing neurons in the same area via reduced inhibitory tone of the NPY/AgRP neurons and a possible contribution of activity changes in orexin/hypocretin expressing neurons in the lateral hypothalamus (14,71). Several recent reviews give excellent overviews on these players, their connectivity, current models and open questions in this fascinating and rapidly advancing field of central appetite and body weight control (67-69,72-74).

In rat adipocytes it has been observed that the GHS-R expression increases with age and during adipogenesis. According to very recently reported data, ghrelin *in vitro* stimulates the differentiation of preadipocytes and antagonizes lipolysis. Ghrelin may therefore play a role in the control of adipogenesis (75). Other effects of ghrelin at the adipocyte level are the suppression of adiponectin expression, and an stimulating influence on mitogen-activated protein kinase phosphorylation (76). Ongoing and future studies are needed to confirm the overall influence of ghrelin on energy homeostasis.

Recent studies emphasize the importance of finding an anti-obesity agent, identifying increased adiposity and its consequences as one of the major killers in western civilizations (77). While one can only speculate based on the currently known scientific facts, if a ghrelin antagonist may represent a useful therapeutic against increased body fat mass, several recent findings may provide arguments for and against a blockade of the ghrelin pathway becoming a treatment option for the rapidly spreading obesity epidemic (78). Apart from representing a possible use as a general anti-obesity drug, a ghrelin-antagonist will help to answer a variety of open questions regarding ghrelin physiology, such as a possible role in meal initiation or its putative involvement in the etiology of PWS (60). As an exception to the generally observed negative correlation between body fat mass and plasma ghrelin concentration, patients with PWS have been identified as the only population of individuals with increased fat mass and several-fold increased plasma ghrelin levels (59). PWS is the most frequent known cause of genetically induced obesity and is associated with a defect on the short arm of chromosome 15, while the exact pathogenetic mechanisms leading to the obesity syndrome in PWS remain unclear (79). Apart from their adiposity, patients with PWS suffer from a severe hunger syndrome that leads to the consumption of non-food items, and they exhibit decreased locomotor activity, impaired growth hormone secretion, increased sleepiness and relative hypoinsulinemia (80,81). Ghrelin, on the other hand, is known to increase fat mass, increase hunger, promote sleep, decrease locomotor activity, control GH-secretion and is suppressed by insulin (11). Although it appears intriguing that hyperghrelinemia in PWS might be responsible, at least in part, for the majority of symptoms characterizing this disease, a genetic link can only be explained via indirect influences (genetic imprinting etc.) since the gene encoding ghrelin is located on chromosome 3 (82). A compound with the ability to neutralize or antagonize ghrelin action could not only prove if ghrelin is causally involved in the pathogenesis of PWS, but also represent the first causal therapeutic approach for a drug treatment of patients with PWS. However, no ghrelin receptor antagonists or ghrelin neutralizing agents are available until now.

The possibly most pressing question concerns the transferability and validity of the above described findings from rodents to humans. Several recent clinical studies on the effects of ghrelin on GH-secretion in humans have reported hunger sensations as the only noticeable side effect in up to 80% of the treated individuals (83,84). Prospective clinical studies focusing on all aspects of energy balance using contemporary methods for the analysis of body composition, energy expenditure, metabolic and endocrine changes can help to clarify these issues. A first clinical study investigating these open questions was conducted by Wren et al., who showed that iv administration of physiologically occurring concentrations of ghrelin effectively triggers appetite and increases food intake in humans (54). Chronic studies investigating the effects of ghrelin and ghrelin receptor agonists are ongoing and will deliver data on long term effects of ghrelin on body composition in humans. Independent from the outcome of these studies, a large number of additional experimental as well as clinical studies will be necessary to determine if a ghrelin antagonist, once it becomes available, may represent an effective and safe treatment of obesity.

It will be important to know which side effects to watch out for during studies attempting to decrease nutrient intake and fat mass by blocking endogenous ghrelin action. Certainly there must be concern about creating growth hormone deficiency, but researchers are even more worried about the antiproliferative, cardiovascular or sleep regulating effects of ghrelin that have been described recently. Possible effects on cardiac rhythm and heart contractility or possible occurrence of malignant diseases in various tissues have to be monitored closely in toxicology studies, even if a ghrelin receptor antagonist is found that has effects on energy balance in rodents. For more extensive information on historical facts, as well as on the biology and the physiology of growth hormone secretagogues (GHS) and ghrelin, numerous comprehensive review articles can be recommended (4,85,86).

Several strategies can be employed to diminish or abolish ghrelin action. Apart from a classical pharmacological approach of antagonizing ghrelin at the ghrelin receptor GHS-R1a (which heavily relies on the notion that this is the only existing or at least the crucial ghrelin receptor), binding or neutralizing ghrelin (e.g. using synthetic antibody fragments) or blocking ghrelin transcription or translation (antisense oligonucleotides) are possible and on the way. A very elegant way of inhibiting ghrelin action would be to block the post-translational acylation process by inactivating the responsible enzyme. Multiple questions have to be clarified on the way to an effective drug using any of the above mentioned strategies. Does the putative drug cross the blood brain barrier or does it even have to? Will it make a difference to further decrease circulating ghrelin levels when obesity is present or will there be resistance to ghrelin antagonists in obesity?

It has became a popular hypothesis among obesity researchers that evolution has shaped an endogenous control system governing body weight and appetite that is based on redundancy to prevent a negative energy balance and to ensure survival ("thrifty gene hypothesis") (87). Although there is no definite proof for this concept, the hypothesis seems intriguing and would make it very unlikely that a ghrelin receptor antagonist alone would cause sustained fat loss, since numerous adjustments of endogenous factors controlling energy balance would occur immediately to keep body weight stable.

On the other hand, there seem to be examples where pharmacological manipulation of circulating hormone levels work well in "cheating" the brain regarding information on physiological functions in the periphery (e.g. oral contraceptives) which are of a comparably essential character for species survival as a stable energy balance (e.g. reproduction). An additional option would be to combine two or three agents to fight obesity. A combination of drugs increasing resting energy expenditure and motivation for locomotor activity as well as decreasing preprandial appetite and increasing post-prandial satiety, might have a higher chance of achieving sustained weight loss, however requires elaborate and costly studies while markedly increasing the risk for side effects. Alternating several drug treatments to avoid resistance (as is historically done e.g., to cure tuberculosis) is another possibility. A more modern strategy may rely on substantial advances in the characterization and diagnosis of specific subtypes of the disease obesity: Based on the hypothesis that there are several molecular reasons to be susceptible to diet-induced obesity (increased ghrelin production or sensitivity, decreased leptin production or sensitivity, decreased melanocortin-4-receptor expression, etc.) effective antiobesity drugs may have to be tailored to meet the specific molecular defects of these sub-groups of patients. This may become possible since both the understanding of mechanisms governing energy balance as well as the clinical and genetic abilities to diagnose obesity phenotypes and genotypes are now rapidly improving (88).

This would have the advantage that one has to worry less about possible side effects of a compound since one would rather aim at correcting a specific defect than trying to modulate physiologically balanced systems. For example, if hyperghrelinemia in PWS patients turns out to be involved in the pathogenesis of the disease, a ghrelin antagonist should theoretically offer a safe and effective treatment option.

In patients with diet-induced obesity, where plasma ghrelin levels are already relatively low and presumably ghrelin does not represent the reason for the disease, one cannot exclude the possibility that ghrelin antagonists might do more harm than good. Since ghrelin apparently displays beneficial hemodynamic effects through both GH-dependent and independent mechanisms, its inactivation could be detrimental for cardiac function. In particular, ghrelin has been shown to reduce cardiac afterload and to increase cardiac output (89), as well as to prevent apoptosis in cardiomyocytes (90), while it is speculated that these effects are not mediated by the ghrelin receptor GHS-R1a, but by the fatty acid scavenger CD36 or other receptors or receptor subtypes (91). As long as the mechanisms and magnitude of ghrelin-induced changes in heart function are poorly understood, there remains the risk of cardiovascular side effects.

The same is the case for (anti)proliferative effects of acylated and non-acylated ghrelin, which have been shown *in vitro* using several tumor cell lines (92). It has been speculated, that these effects were based on the activation of an until now unknown GHS receptor (93). In contrast, other findings seem to support the concept that at least in some organs such as prostate and adrenal gland, ghrelin may represent a proliferative stimulus through activation of the GHS1a receptor (94). Furthermore, ghrelin has been shown to be a sleep promoting factor (95). A putative ghrelin antagonist may thererefore impair sleep and have a negative influence on bone formation and remodeling. It is difficult to predict the action of a new compound having ghrelinergic action or antagonistic properties without knowing its affinity and specificity to putatively existing types and subtypes of ghrelin receptors. Possible oncogenic, cardiovascular, sleep-related and bone-density affecting side effects should be monitored carefully among others, when these compounds will be tested.

While obesity clearly represents a rapidly spreading disease with an enormous market potential due to what will be most likely life long drug treatment, ghrelin receptor agonists are not only already available as orally active compounds, but also could be more effective as a drug for cachexia than ghrelin receptor antagonists for obesity, due to a less potent defensive system protecting against body weight increase. Ghrelin receptor agonists or ghrelin might therefore offer drug treatment for diseases such as cancer cachexia, HIV-wasting syndrome, cardiac cachexia or even anorexia nervosa. While comparable with leptin resistance in obesity, cachectic patients might be resistant to treatment with ghrelin since their endogenous ghrelin levels are markedly increased; preliminary results in rodents bearing melanoma cells show encouraging results (96). Again, side effects might occur: in particular GH-mediated stimulation of IGF-I might be an undesired effect in at least some malignant diseases (97). In addition, while ghrelin or its receptor agonists may promote appetite and food intake, it remains to be shown that these agents can increase body weight and fat mass in humans. In rodents adipogenic effects are most potent during the first 14 days of (peripheral) treatment and may disappear due to desensitization after a few weeks.

In summary, ghrelin represents a recently discovered gastric hormone that induces hunger and increased fat deposition via central and possibly additional peripheral mechanisms in response to a negative energy balance. Ghrelin is one of the most potent and the only known peripheral orexigenic agent, and it possibly represents the first meal initiation factor. Although plasma ghrelin concentrations are negatively correlated with fat mass, substantial levels are still secreted in the vast majority of obese individuals, while in patients with a gastric bypass weight loss occurs along with loss of circulating ghrelin. Apart from a possible effectiveness of a ghrelin antagonist for the general prophylaxis and treatment of adiposity, the blockade of ghrelin could be the first specific pharmaco-therapeutic approach to successfully treat patients with PWS. However, various additional effects of ghrelin on physiological processes and organ systems implicate not only therapeutic perspectives, but make unwanted cardiovascular, gastrointestinal or proliferative effects caused by the blockade of ghrelin action a likely occurrence.

REFERENCES

1. Kojima M, Hosoda H, Date Y, Nakazato M, Matsuo H, Kangawa K. Ghrelin is a growth-hormone-releasing acylated peptide from stomach. Nature. 1999; 402:656-60.
2. Zhang Y, Proenca R, Maffei M, Barone M, Leopold L, Friedman JM. Positional cloning of the mouse obese gene and its human homologue. Nature. 1994; 372:425-32.
3. Howard AD, Feighner SD, Cully DF, et al. A receptor in pituitary and hypothalamus that functions in growth hormone release. Science. 1996; 273:974-7.
4. Bowers CY. Unnatural growth hormone-releasing peptide begets natural ghrelin. J Clin Endocrinol Metab. 2001; 86:1464-9.
5. Crowley VE, Yeo GS, O'Rahilly S. Obesity therapy: altering the energy intake-and-expenditure balance sheet. Nat Rev Drug Discov. 2002; 1:276-86.
6. Marx J. Cellular warriors at the battle of the bulge. Science. 2003; 299:846-9.
7. Hosoda H, Kojima M, Matsuo H, Kangawa K. Ghrelin and des-acyl ghrelin: two major forms of rat ghrelin peptide in gastrointestinal tissue. Biochem Biophys Res Commun. 2000; 279:909-13.
8. Bednarek MA, Feighner SD, Pong SS, et al. Structure-function studies on the new growth hormone-releasing peptide, ghrelin: minimal sequence of ghrelin necessary for activation of growth hormone secretagogue receptor 1a. J Med Chem. 2000; 43:4370-6.
9. Hosoda H, Kojima M, Mizushima T, Shimizu S, Kangawa K. Structural divergence of human ghrelin. Identification of multiple ghrelin-derived molecules produced by post-translational processing. J Biol Chem. 2003; 278:64-70.
10. Beaumont NJ, Skinner VO, Tan TM, et al. Ghrelin can bind to a species of high density lipoprotein associated with paraoxonase. J Biol Chem. 2003; 278:8877-80.
11. Muccioli G, Tschöp M, Papotti M, Deghenghi R, Heiman M, Ghigo E. Neuroendocrine and peripheral activities of ghrelin: implications in metabolism and obesity. Eur J Pharmacol. 2002; 440:235-54.
12. Date Y, Kojima M, Hosoda H, et al. Ghrelin, a novel growth hormone-releasing acylated peptide, is synthesized in a distinct endocrine cell type in the gastrointestinal tracts of rats and humans. Endocrinology. 2000; 141:4255-61.
13. Yoshihara F, Kojima M, Hosoda H, Nakazato M, Kangawa K. Ghrelin: a novel peptide for growth hormone release and feeding regulation. Curr Opin Clin Nutr Metab Care. 2002; 5:391-5.
14. Cowley MA, Smith RG, Diano S, et al. The distribution and mechanism of action of ghrelin in the CNS demonstrates a novel hypothalamic circuit regulating energy homeostasis. Neuron. 2003; 37:649-61.
15. Ariyasu H, Takaya K, Tagami T, et al. Stomach is a major source of circulating ghrelin, and feeding state determines plasma ghrelin-like immunoreactivity levels in humans. J Clin Endocrinol Metab. 2001; 86:4753-8.

16. Date Y, Murakami N, Toshinai K, et al. The role of the gastric afferent vagal nerve in ghrelin-induced feeding and growth hormone secretion in rats. Gastroenterology. 2002; 123:1120-8.

17. Tschöp M, Smiley DL, Heiman ML. Ghrelin induces adiposity in rodents. Nature. 2000; 407:908-13.

18. Lall S, Tung LY, Ohlsson C, Jansson JO, Dickson SL. Growth hormone (GH)-independent stimulation of adiposity by GH secretagogues. Biochem Biophys Res Commun. 2001; 280:132-8.

19. Tschöp M, Statnick MA, Suter TM, Heiman ML. GH-releasing peptide-2 increases fat mass in mice lacking NPY: indication for a crucial mediating role of hypothalamic agouti-related protein. Endocrinology. 2002; 143:558-68.

20. Nakazato M, Murakami N, Date Y, et al. A role for ghrelin in the central regulation of feeding. Nature. 2001; 409:194-8.

21. Shuto Y, Shibasaki T, Otagiri A, et al. Hypothalamic growth hormone secretagogue receptor regulates growth hormone secretion, feeding, and adiposity. J Clin Invest. 2002; 109:1429-36.

22. Horvath TL, Diano S, Sotonyi P, Heiman M, Tschöp M. Minireview: ghrelin and the regulation of energy balance--a hypothalamic perspective. Endocrinology. 2001; 142:4163-9.

23. Wren AM, Small CJ, Ward HL, et al. The novel hypothalamic peptide ghrelin stimulates food intake and growth hormone secretion. Endocrinology. 2000; 141:4325-8.

24. Matsumoto M, Hosoda H, Kitajima Y, et al. Structure-activity relationship of ghrelin: pharmacological study of ghrelin peptides. Biochem Biophys Res Commun. 2001; 287:142-6.

25. Ariyasu H, Takaya K, Hosoda H, et al. Delayed Short-Term Secretory Regulation of Ghrelin in Obese Animals: Evidenced by a Specific RIA for the Active Form of Ghrelin. Endocrinology. 2002; 143:3341-50.

26. Rindi G, Necchi V, Savio A, et al. Characterisation of gastric ghrelin cells in man and other mammals: studies in adult and fetal tissues. Histochem Cell Biol. 202; 117:511-9.

27. Folwaczny C, Chang JK, Tschöp M. Ghrelin and motilin: two sides of one coin? Eur J Endocrinol. 2001; 144:R1-R3.

28. Havel PJ. Peripheral signals conveying metabolic information to the brain: short-term and long-term regulation of food intake and energy homeostasis. Exp Biol Med. 2001; 226:963-77.

29. Toshinai K, Mondal MS, Nakazato M, et al. Upregulation of Ghrelin expression in the stomach upon fasting, insulin-induced hypoglycemia, and leptin administration. Biochem Biophys Res Commun. 2001; 281:1220-5.

30. Cummings DE, Weigle DS, Frayo RS, et al. Plasma ghrelin levels after diet-induced weight loss or gastric bypass surgery. N Engl J Med. 2002; 346:1623-30.

31. Gualillo O, Caminos J, Blanco M, et al. Ghrelin, a novel placental-derived hormone. Endocrinology. 2001; 142:788-94.

32. Volante M, AllIa E, Gugliotta P, et al. Expression of ghrelin and of the GH secretagogue receptor by pancreatic islet cells and related endocrine tumors. J Clin Endocrinol Metab. 2002; 87:1300-8.

33. Wierup N, Svensson H, Mulder H, Sundler F. The ghrelin cell: a novel developmentally regulated islet cell in the human pancreas. Regul Pept. 2002; 107:63-9.

34. Korbonits M, Bustin SA, Kojima M, et al. The expression of the growth hormone secretagogue receptor ligand ghrelin in normal and abnormal human pituitary and other neuroendocrine tumors. J Clin Endocrinol Metab. 2001; 86:881-7.

35. Gnanapavan S, Kola B, Bustin SA, et al. The tissue distribution of the mRNA of ghrelin and subtypes of its receptor, GHS-R, in humans. J Clin Endocrinol Metab. 2002; 87:2988-91.

36. Mori K, Yoshimoto A, Takaya K, et al. Kidney produces a novel acylated peptide, ghrelin. FEBS Lett. 2000; 486:213-6.

37. Hattori N, Saito T, Yagyu T, Jiang BH, Kitagawa K, Inagaki C. GH, GH receptor, GH secretagogue receptor, and ghrelin expression in human T cells, B cells, and neutrophils. J Clin Endocrinol Metab. 2001; 86:4284-91.

38. Broglio F, Gottero C, Arvat E, Ghigo E. Endocrine and non-endocrine actions of ghrelin. Horm Res. 2003; 59:109-17.

39. Kojima M, Hosoda H, Kangawa K Purification and distribution of ghrelin: the natural endogenous ligand for the growth hormone secretagogue receptor. Horm Res. 2001; 56:93-7.

40. Cummings DE, Purnell JQ, Frayo RS, Schmidova K, Wisse BE, Weigle DS. A preprandial rise in plasma ghrelin levels suggests a role in meal initiation in humans. Diabetes. 2001; 50:1714-9.

41. Tschöp M, Wawarta R, Riepl RL, et al. Post-prandial decrease of circulating human ghrelin levels. J Endocrinol Invest. 2001; 24:RC19-21.

42. Hayashida T, Murakami K, Mogi K, et al. Ghrelin in domestic animals: distribution in stomach and its possible role. Domest Anim Endocrinol. 2001; 21:17-24.

43. Dube MG, Beretta E, Dhillon H, Ueno N, Kalra PS, Kalra SP. Central leptin gene therapy blocks high-fat diet-induced weight gain, hyperleptinemia, and hyperinsulinemia: increase in serum ghrelin levels. Diabetes. 2002; 51:1729-36.

44. Caixas A, Bashore C, Nash W, Pi-Sunyer F, Laferrere B. Insulin, unlike food intake, does not suppress ghrelin in human subjects. J Clin Endocrinol Metab. 2002; 87:1902-6.

45. McCowen KC, Maykel JA, Bistrian BR, Ling PR. Circulating ghrelin concentrations are lowered by intravenous glucose or hyperinsulinemic euglycemic conditions in rodents. J Endocrinol. 2002; 175:R7-11.

46. Mohlig M, Spranger J, Otto B, Ristow M, Tschöp M, Pfeiffer AF 2002. Euglycemic hyperinsulinemia, but not lipid infusion, decreases circulating ghrelin levels in humans. J Endocrinol Invest 25:RC36-8.

47. Saad MF, Bernaba B, Hwu CM, et al. Insulin regulates plasma ghrelin concentration. J Clin Endocrinol Metab. 2002; 87:3997-4000.

48. Flanagan DE, Evans ML, Monsod TP, et al. The influence of insulin on circulating ghrelin. Am J Physiol Endocrinol Metab. 2003; 284:E313-6.

49. Schaller G, Schmidt A, Pleiner J, Woloszczuk W, Wolzt M, Luger A. Plasma ghrelin concentrations are not regulated by glucose or insulin: a double-blind, placebo-controlled crossover clamp study. Diabetes. 2003; 52:16-20.

50. Janssen JA, van der Toorn FM, Hofland LJ, et al. Systemic ghrelin levels in subjects with growth hormone deficiency are not modified by one year of growth hormone replacement therapy. Eur J Endocrinol. 2001; 145:711-6.

51. Tschöp M, Flora DB, Mayer JP, Heiman ML. Hypophysectomy prevents ghrelin-induced adiposity and increases gastric ghrelin secretion in rats. Obes Res. 2002; 10:991-9.

52. Cappiello V, Ronchi C, Morpurgo PS, et al. Circulating ghrelin levels in basal conditions and during glucose tolerance test in acromegalic patients. Eur J Endocrinol. 2002; 147:189-94.

53. Papotti M, Cassoni P, Volante M, Deghenghi R, Muccioli G, Ghigo E. Ghrelin-producing endocrine tumors of the stomach and intestine. J Clin Endocrinol Metab. 2001; 86:5052-9.

54. Wren AM, Seal LJ, Cohen MA, et al. Ghrelin enhances appetite and increases food intake in humans. J Clin Endocrinol Metab. 2001; 86:5992-5.

55. Tschöp M, Weyer C, Tataranni PA, Devanarayan V, Ravussin E, Heiman ML. Circulating ghrelin levels are decreased in human obesity. Diabetes. 2001; 50:707-9.
56. Korbonits M, Gueorguiev M, O'Grady E, et al. A variation in the ghrelin gene increases weight and decreases insulin secretion in tall, obese children. J Clin Endocrinol Metab. 2002; 87:4005-8.
57. Hinney A, Hoch A, Geller F, et al. Ghrelin gene: identification of missense variants and a frameshift mutation in extremely obese children and adolescents and healthy normal weight students. J Clin Endocrinol Metab. 2002; 87:2716-9.
58. Ukkola O, Ravussin E, Jacobson P, et al. Role of ghrelin polymorphisms in obesity based on three different studies. Obes Res. 2002; 10:782-91.
59. Cummings DE, Clement K, Purnell JQ, et al. Elevated plasma ghrelin levels in Prader Willi syndrome. Nat Med. 2002; 8:643-4.
60. DelParigi A, Tschöp M, Heiman ML, et al. High circulating ghrelin: a potential cause for hyperphagia and obesity in prader-willi syndrome. J Clin Endocrinol Metab. 2002; 87:5461-4.
61. Haqq AM, Farooqi IS, O'Rahilly S, et al. Serum ghrelin levels are inversely correlated with body mass index, age, and insulin concentrations in normal children and are markedly increased in Prader-Willi syndrome. J Clin Endocrinol Metab. 2003; 88:174-8.
62. Otto B, Cuntz U, Fruehauf E, et al. Weight gain decreases elevated plasma ghrelin concentrations of patients with anorexia nervosa. Eur J Endocrinol. 2001; 145:669-73.
63. Hewson AK, Tung LY, Connell DW, Tookman L, Dickson SL. The rat arcuate nucleus integrates peripheral signals provided by leptin, insulin, and a ghrelin mimetic. Diabetes. 2002; 51:3412-9.
64. Wren AM, Small CJ, Abbott CR, et al. Ghrelin causes hyperphagia and obesity in rats. Diabetes. 2001; 50:2540-7.
65. Torsello A, Locatelli V, Melis MR, et al. Differential orexigenic effects of hexarelin and its analogs in the rat hypothalamus: indication for multiple growth hormone secretagogue receptor subtypes. Neuroendocrinology. 2000; 72:327-32.
66. Jeanrenaud B, Rohner-Jeanrenaud F. Effects of neuropeptides and leptin on nutrient partitioning: dysregulations in obesity. Annu Rev Med. 2001; 52:339-51.
67. Berthoud HR. Multiple neural systems controlling food intake and body weight. Neurosci Biobehav Rev. 2002; 26:393-428.
68. Saper CB, Chou TC, Elmquist JK. The need to feed: homeostatic and hedonic control of eating. Neuron. 2002; 36:199-211.
69. Grill HJ, Kaplan JM. The neuroanatomical axis for control of energy balance. Front Neuroendocrinol. 2002; 23:2-40.
70. Willesen MG, Kristensen P, Romer J. Co-localization of growth hormone secretagogue receptor and NPY mRNA in the arcuate nucleus of the rat. Neuroendocrinology. 1999; 70:306-16.
71. Toshinai K, Date Y, Murakami N, et al. Ghrelin-induced food intake is mediated via the orexin pathway. Endocrinology. 2003; 144:1506-12.
72. Kalra SP, Dube MG, Pu S, Xu B, Horvath TL, Kalra PS. Interacting appetite-regulating pathways in the hypothalamic regulation of body weight. Endocr Rev. 1999; 20:68-100.
73. Schwartz MW, Woods SC, Porte D, Jr., Seeley RJ, Baskin DG. Central nervous system control of food intake. Nature. 2000; 404:661-71.
74. Spiegelman BM, Flier JS. Obesity and the regulation of energy balance. Cell. 2001; 104:531-43.
75. Choi K, Roh SG, Hong YH, et al. The role of ghrelin and growth hormone secretagogues receptor in rat adipogenesis. Endocrinology. 2003; 144:754-9.

76. Ott V, Fasshauer M, Dalski A, et al. Direct peripheral effects of ghrelin include suppression of adiponectin expression. Horm Metab Res. 2002; 34:640-5.

77. Mokdad AH, Ford ES, Bowman BA, et al. Prevalence of obesity, diabetes, and obesity-related health risk factors, 2001. JAMA. 2003; 289:76-9.

78. Flier JS, Maratos-Flier E. The stomach speaks--ghrelin and weight regulation. N Engl J Med. 2002; 346:1662-3.

79. Nativio DG. The genetics, diagnosis, and management of Prader-Willi syndrome. J Pediatr Health Care. 2002; 16:298-303.

80. Burman P, Ritzen EM, Lindgren AC. Endocrine dysfunction in Prader-Willi syndrome: a review with special reference to GH. Endocr Rev. 2001; 22:787-99.

81. Eiholzer U, Nordmann Y, l'Allemand D, Schlumpf M, Schmid S, Kromeyer-Hauschild K. Improving body composition and physical activity in Prader-Willi Syndrome. J Pediatr. 2003; 142:73-8.

82. Scott AF, 2000 [http://www.ncbi.nlm.nih.gov/omim/, Online Mendelian Inheritance in Men (MIM no. 605353); Scott, A. F., personal communication in OMIM].

83. Arvat E, Di Vito L, Broglio F, et al. Preliminary evidence that Ghrelin, the natural GH secretagogue (GHS)-receptor ligand, strongly stimulates GH secretion in humans. J Endocrinol Invest. 2000; 23:493-5.

84. Broglio F, Arvat E, Benso A, et al. Ghrelin, a natural GH secretagogue produced by the stomach, induces hyperglycemia and reduces insulin secretion in humans. J Clin Endocrinol Metab. 2001; 86:5083-6.

85. Giustina A, Veldhuis JD. Pathophysiology of the neuroregulation of growth hormone secretion in experimental animals and the human. Endocr Rev. 1998; 19:717-97.

86. van der Lely AJ, Tschöp M, Heiman ML, Ghigo E. Biological, physiological, pathophysiological and pharmacological aspects of ghrelin. Endocrine Reviews. 2004; *in press*.

87. Bray GA. Hypothalamic and genetic obesity: an appraisal of the autonomic hypothesis and the endocrine hypothesis. Int J Obes. 1984; 8:119-37.

88. Farooqi IS, Keogh JM, Yeo GS, Lank EJ, Cheetham T, O'Rahilly S. Clinical spectrum of obesity and mutations in the melanocortin 4 receptor gene. N Engl J Med. 2003; 348:1085-95.

89. Nagaya N, Miyatake K, Uematsu M, et al. Hemodynamic, renal, and hormonal effects of ghrelin infusion in patients with chronic heart failure. J Clin Endocrinol Metab. 2001; 86:5854-9.

90. Baldanzi G, Filigheddu N, Cutrupi S, et al. Ghrelin and des-acyl ghrelin inhibit cell death in cardiomyocytes and endothelial cells through ERK1/2 and PI 3-kinase/AKT. J Cell Biol. 2002; 159:1029-37.

91. Bodart V, Febbraio M, Demers A, et al. CD36 mediates the cardiovascular action of growth hormone-releasing peptides in the heart. Circ Res. 2002; 90:844-9.

92. Ghè C, Cassoni P, Catapano F, et al. The antiproliferative effect of synthetic peptidyl GH secretagogues in human CALU-1 lung carcinoma cells. Endocrinology. 2002; 143:484-91.

93. Cassoni P, Papotti M, Ghe C, et al. Identification, characterization, and biological activity of specific receptors for natural (ghrelin) and synthetic growth hormone secretagogues and analogs in human breast carcinomas and cell lines. J Clin Endocrinol Metab. 2001; 86:1738-45.

94. Jeffery PL, Herington AC, Chopin LK. The potential autocrine/paracrine roles of ghrelin and its receptor in hormone-dependent cancer. Cytokine Growth Factor Rev. 2003; 14:113-22.

95. Weikel JC, Wichniak A, Ising M, et al. Ghrelin promotes slow-wave sleep in humans. Am J Physiol Endocrinol Metab. 2003; 284:E407-15.

96. Hanada T, Toshinai K, Kajimura N, et al. Anti-cachectic effect of ghrelin in nude mice bearing human melanoma cells. Biochem Biophys Res Commun. 2003; 301:275-9.
97. Cohen P, Clemmons DR, Rosenfeld RG. Does the GH-IGF axis play a role in cancer pathogenesis? Growth Horm IGF Res. 2000; 10:297-305.

Chapter 8

CARDIOVASCULAR ACTIVITIES OF GHRELIN AND SYNTHETIC GHS

Jörgen Isgaard, Inger Johansson & Åsa Tivesten
Research Center for Endocrinology and Metabolism, Sahlgrenska University Hospital, Göteborg, Sweden

Abstract: Recent experimental data demonstrate cardiovascular effects of synthetic growth hormone secretagogues (GHS). These cardiovascular effects include improvement of systolic function in rats after experimental infarction, cardioprotection against postischemic dysfunction in perfused rat hearts and increase of left ventricular ejection fraction in hypopituitary patients. The proposed natural ligand ghrelin has been isolated and characterized from rat stomach. It was recently reported that a single injection of ghrelin to healthy volunteers decreased blood pressure and increased stroke volume and cardiac output. Similar beneficial cardiovascular effects were observed when ghrelin was administered to patients with chronic heart failure and to rats with experimental myocardial infarction. Specific binding of GHS to rat cardiac membranes and human cardiac tissue has been reported and possible growth hormone (GH) independent effects of GHS have been suggested. We have recently used H9c2 cardiac cells to demonstrate specific and dose-dependent stimulation of thymidine incorporation by GHS and ghrelin. Moreover, binding studies on H9c2 cells demonstrate specific binding of GHS and ghrelin and add further support for an alternative subtype-binding site in the heart compared to the pituitary. In conclusion, accumulating data suggest beneficial effects of GHS and ghrelin on cardiovascular function and these effects may be at least partly independent of GH.

Key words: growth hormone secretagogues, hexarelin, heart, cardiovascular

1.   INTRODUCTION

In addition to their metabolic effects, both synthetic growth hormone secretagogues (GHS) and ghrelin have clear effects on the cardiovascular system.

In this chapter, we will give an overview regarding reported cardioprotective effects against ischemia and other effects in cardiovascular disease. Moreover, GHS and ghrelin effects on cardiomyocytes and their role in cardiovascular physiology will be discussed.

An intriguing aspect of the mechanisms of GHS and ghrelin that will be discussed is that some of the cardiovascular effects may be direct on the heart and vasculature independent of growth hormone (GH).

2.   CARDIOPROTECTION AND OTHER EFFECTS IN CARDIOVASCULAR DISEASE

An accumulating number of both experimental and clinical studies have shown a variety of beneficial cardiovascular effects of GH and insulin-like growth factor-I (IGF-I) in different settings of impaired cardiac function (1-3). This has prompted several research groups to study possible cardioprotective effects of GHS and ghrelin during conditions of ischemia and impaired cardiac function.

In a study by De Gennaro Colonna and collaborators, antiserum to growth hormone releasing hormone (GHRH) was used to achieve GH deficiency in rats prior to treatment with GH or the synthetic peptidyl GHS hexarelin for 2 weeks (4). After killing the rats their hearts were subjected to retrograde aortic perfusion under ischemic conditions. Control rats with GH deficiency showed marked increase in left ventricular (LV) end diastolic pressure and poor recovery of contractility after reperfusion. In contrast, rats treated with GH or hexarelin normalized cardiac function to a similar extent. In a more recent study, hypophysectomized rats were used in a similar experimental setting and again pretreatment with hexarelin and GH was protective against ischemic damage during subsequent perfusion (5).

Interestingly, both these studies suggest GH independent effects since hexarelin had effects on cardiac peformance in two different models of GH deficiency in rats.

Temporary myocardial ischemia followed by reperfusion causes reversible cardiac dysfunction known as myocardial stunning. A rabbit model was recently used to study effects of GH and GHS on myocardial stunning (6). Animals were pretreated with GH or Growth Hormone Releasing Peptide (GHRP)-2 for two weeks. Subsequently, hearts were blood perfused and subjected to 15 minutes of ischemia followed by 80 minutes of reperfusion and compared with nonischemic hearts. There was no difference in post-ischemic recovery of LV systolic function in any group. However, there was a significant decrease of LV end diastolic pressure by GHRP-2 after reperfusion, suggesting a beneficial effect on diastolic function in this animal model.

In a recent study from our laboratory, a different experimental approach was used (7). Intact rats were subjected to experimental myocardial infarction and after 4 weeks of recovery, the rats were treated with two doses of hexarelin (10 or 100 μg/kg per day), GH 2.5 mg/kg per day or saline. Cardiac structure and function were evaluated with echocardiography at baseline and at the end of the experiment. Stroke volume was significantly increased in both the high dose hexarelin group and GH group compared to control rats. Similar findings were also seen regarding cardiac output, which was also increased by hexarelin and GH. Both these parameters remained significantly elevated also when normalized to body weight. Interestingly, only GH treatment significantly increased body weight and kidney IGF-I mRNA that may suggest the effects of hexarelin were independent of GH.

As an alternative method to coronary artery ligation, electric pacing of the heart is sometimes used as a model for impaired cardiac function. King and collaborators induced rapid pacing at 240 bpm for 3 weeks in pigs (8). An orally bioavailable GHS called CP-424,391 was administered to one group of animals at initiation of pacing. After 3 weeks of pacing and treatment, GHS treated pigs had significantly higher fractional shortening and lower wall stress due to cardiac hypertrophy compared to control animals. Serum IGF-I was increased approximately two-fold which would suggest an effect at least partially mediated by an increased GH secretion.

The first study using ghrelin, the endogenous ligand for the GHS receptor (GHS-R), to study effects after myocardial injury, was recently published. Kangawa and collaborators used rats with experimental myocardial infarction and treated them with ghrelin 100 μg daily for three weeks (9). Echocardiography and catheterization were used to evaluate effects of ghrelin, showing higher cardiac output and fractional shortening in ghrelin treated rats compared to controls. A clinical trial with acute administration of ghrelin to patients with congestive heart failure has also been published by the same investigators (10). Twelve patients with chronic hearth failure were given a single intavenous infusion of human ghrelin or placebo. Ghrelin significantly increased cardiac index, stroke volume index and decreased systemic vascular resistance within 60 minutes. Moreover, in anesthetized patients undergoing bypass surgery, hexarelin was found to increase ejection fraction, cardiac output and mean arterial pressure without any changes in peripheral resistance (11).

In a recent *in vitro* study, it was reported that both synthetic GHS and ghrelin inhibit apoptosis of primary adult guinea pig and H9c2 cardiomyocytes and endothelial cells through binding to a receptor distinct from the GHS-R type 1a (12). The proposed signaling was shown to include activation of ERK 1/2 and Akt serine kinases.

To summarize, accumulating evidence suggest beneficial effects of GHS on both diastolic and systolic function after ischemic injury and this has been shown in several different experimental models and a recent clinical trial with chronic heart failure patients.

Moreover, some of these effects are present also in the absence of GH, suggesting direct interaction between GHS and the cardiovascular system.

3.   GHS AND GHRELIN EFFECTS ON CARDIOMYOCYTES

Supporting evidence for possible GH independent effects by GHS was provided when it was reported that an iodinated and photoactivable derivative of hexarelin could specifically bind to rat cardiac membranes (13,14). Binding of hexarelin to H9c2 cardiomyocytes (15) and human cardiac tissues (16) has also been reported.

To further explore the possibility of direct effects of GHS on the heart we have used rat cardiomyocyte H9c2 cells to study if cell proliferation could be stimulated by hexarelin. As a marker for DNA synthesis and proliferation we used thymidine incorporation (17). The H9c2 cells are derived from rat ventricle and have many cardiomyocyte properties including IGF-I responsiveness (18-20).

Cells were washed, cultured overnight without fetal calf serum and stimulated with hexarelin. A concentration of 10 µM had a significant effect on thymidine incorporation after 12 hours with a maximal effect after 18 hours. There was a dose-response relationship regarding hexarelin stimulation of thymidine incorporation with significant effects at 3 µM and maximal effect at around 30 µM.

To test the specificity, we looked at different GHS at 10 and 100 µM concentrations after 18 hours of stimulation. Tyr-ala-hexarelin is an hexarelin derivate used in binding assays. EP80317 is a GHRP antagonist, which does not release GH but has a strong affinity for the GHS-R. MK-0677 is a non-peptide GHS with strong GH-releasing properties but does not bind to cardiac GHS-R according to Ong and coworkers (21). EP51389 is a truncated GHRP with strong GH releasing properties but does not bind to cardiac cells. Our results demonstrate similar stimulatory effects by hexarelin, Tyr-ala-hexarelin and EP80317 on thymidine incorporation. There was no stimulatory effect by MK-0677 and EP51389. In fact, we saw a significant inhibitory effect of these GHS on thymidine incorporation. The reason for this is not clear.

Ghrelin also stimulated thymidine incorporation in a dose-dependent manner, although the effect was somewhat weaker than that seen with hexarelin (Figure 8-1).

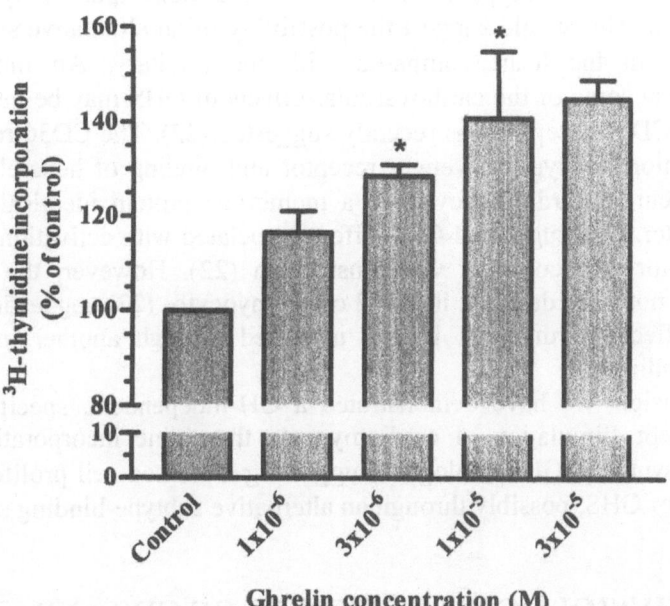

Figure 8-1. Effect of ghrelin on 3H-thymidine incorporation in H9c2 cells. Cells were pretreated and incubated with indicated concentrations of ghrelin. After 18 hours, the cells were harvested and the radioactivity was measured. Data is presented as mean ± standard error of the mean from one representative experiment in triplicate. * $p<0.05$ vs vehicle control.

H9c2 cell membranes were used for the binding studies and the specific binding of ^{125}I-labeled Tyr-Ala-hexarelin was calculated as the difference between binding in the absence (total binding) and in the presence of 10 μM unlabeled Tyr-Ala-hexarelin (nonspecific binding). Experiments using increasing concentrations of ^{125}I-labeled Tyr-Ala-hexarelin revealed the presence of a saturable specific binding associated with a high nonspecific binding that linearly increased as a function of the radioligand concentrations. Scatchard analysis of the specific binding data suggested the existence of a single class of binding sites with a dissociation constant (K_d)

of 7.5±1.0nM and a maximal binding capacity (B_{max}) of 2023±168 fmol/mg of protein (mean±standard error of the mean of three separate experiments). Tyr-Ala-hexarelin, hexarelin, EP80317 and human ghrelin were able to displace the radioligand Tyr-Ala-hexarelin from binding sites in a dose-dependent manner. In contrast, no or little competition was observed in the presence of other competitors such as MK-0677, EP51389 or somatostatin-14. Interestingly, no expression of the GHS-R could be detected using amplification of GHS-R type1a mRNA by reverse transcriptase-polymerase chain reaction. This would suggest the possibility of an alternative subtype-binding site in the heart compared with the pituitary. An intriguing possibility that some of the cardiovascular effects of GHS may be mediated through the CD36 receptor was recently suggested (22). The CD36 receptor is multifunctional B-type scavenger receptor and binding of hexarelin was found to occur in cardiomyocytes to a membrane protein identical to rat CD36 receptor. One suggested GHS effect associated with activation of the CD36 receptor was coronary vasoconstriction (22). However, the CD36 receptor has not been detected in H9c2 cardiomyocytes (23) suggesting that our GHS effects in this cell line is mediated through another, not yet identified binding site.

In conclusion, we have demonstrated a GH-independent, specific and dose-dependent stimulation of cardiomyocyte thymidine incorporation by natural and synthetic GHS analogues suggesting increased cell proliferation and binding of GHS, possibly through an alternative subtype-binding site.

4.   PHYSIOLOGICAL EFFECTS OF GHS AND GHRELIN ON THE CARDIOVASCULAR SYSTEM

Little so far has been reported regarding physiological hemodynamic effects of ghrelin and synthetic GHS. However, it was recently reported that a single injection of hexarelin increased LV ejection fraction in both healthy male volunteers and hypopituitary patients (24,25). Interestingly, peripheral resistance was not affected in these studies, which would suggest myocardial effects behind the increased contractility rather than effects on the vasculature. However, a more recent study (26) where healthy volunteers received a single injection of ghrelin showed a significant decrease in mean arterial pressure and a significant increase in stroke volume index and cardiac index without affecting heart rate.

Possible effects of ghrelin on the vasculature were recently supported by Wiley and Davenport (27) who demonstrated that ghrelin is a potent physiological antagonist of endothelin-1. In a study by Rossoni and coworkers (28) pre-treatment with hexarelin for one week antagonized the ex vivo hyper responsiveness of aortic tissue resulting from hypophysectomy and allowed recovery of acetylcholine responsiveness and generation of prostaglandin I_2 metabolites in the absence of GH. It was also demonstrated in the same study that pretreatment with hexarelin increased the tension response to a nitric oxide (NO)-synthase inhibitor, suggesting that hexarelin enhances NO synthase activity *in vivo*. On the other hand, locally administered ghrelin has an acute vasodilating effect in humans which was not counteracted by an NO-synthase inhibitor (29), suggesting independence of NO production. Taken together, these data would suggest that both ghrelin and synthetic GHS when given as a nonacute treatment have vascular effects and that some of these effects may be beneficial for endothelial function independent of GH.

In conclusion, there is an increasing amount of evidence that both synthetic and natural GHS have cardioprotective effects after ischemic injury. GHS binds specifically to cardiomyocyte membranes and stimulates thymidine incorporation. Moreover, ghrelin act as a vasodilator and there are also possible direct myocardial effects of GHS enhancing contractility. Taken together, these results offer an interesting perspective on the future where further studies aiming at evaluating a role of GHS and ghrelin in the treatment of cardiovascular disease are warranted.

REFERENCES

1. Yang R, Bunting S, Gillett N, Clark R, Jin H. Growth hormone improves cardiac performance in experimental performance. Circulation. 1995; 92:262-7.
2. Tivesten Å, Caidahl K, Kujacic V et al. Similar cardiovascular effects of growth hormone and insulin-like growth factor-I in rats after experimental myocardial infarction. Growth Hormone & IGF Res. 2001; 11:187-95.
3. Fazio S, Sabatino D, Capaldo B et al. A preliminary study of growth hormone in the treatment of dilated cardiomyopathy. New Engl J Med. 1996; 334:809-14.
4. De Gennaro Colonna V, Rossoni G, Bernareggi M, Müller EE, Berti F. Cardiac ischemia and impairment of vascular endothelium function in hearts from growth hormone deficient rats: protection by hexarelin. Eur J Pharm. 1997; 334:201-7.
5. Locatelli V, Rossoni G, Schweiger F et al. Growth hormone-independent cardioprotective effects of hexarelin in the rat. Endocrinology. 1999; 140:4024-31.
6. Weekers F, Van Herck E, Isgaard J, Van den Berghe G. Pretreatment with growth hormone-releasing peptide-2 directly protects against the diastolic dysfunction of myocardial stunning in an isolated, blood perfused rabbit model. Endocrinology. 2000; 141:3993-9.
7. Tivesten Å, Bollano E, Caidahl K et al. The growth hormone secretagogue hexarelin improves cardiac function in rats after experimental myocardial infarction. Endocrinology. 2000; 141:60-6.

8. King MK, Gay DM, Pan LC. Treatment with a growth hormone secretagogue in a model of developing heart failure. Effects on ventricular function. Circulation. 2001; 103:308-13.

9. Nagaya N, Uematsu M, Kojima M et al. Chronic administration of ghrelin improves left ventricular dysfunction and attenuates development of cardiac cachexia in rats with heart failure. Circulation. 2001; 104:1430-5.

10. Nagaya N, Miyatake K, Uematsu M et al. Hemodynamic, renal and hormonal effects of ghrelin infusion in patients with chronic heart failure. J Clin Endocrinol Metab. 2001; 86:5854-9.

11. Broglio F, Guarracino F, Benso A et al. Effects of acute hexarelin administration on cardiac performance in patients with coronary artery disease during by-pass surgery. Eur J Pharmacol. 2002; 448:193-200.

12. Baldanzi G, Filigheddu N, Cutrupi S et al. Ghrelin and des-acyl ghrelin inhibit cell death in cardiomyocytes and endothelial cells through ERK1/2 and PI 3-kinase/AKT. J Cell Biol. 2002; 159:1029-37.

13. Ong H, McNicoll N, Escher E et al. Identification of a pituitary growth hormone-releasing peptide (GHRP) receptor subtype by photoaffinity labeling. Endocrinology. 1998; 139:432-5.

14. Bodart V, Bouchard JF, McNicoll N et al. Identification and characterization of a new growth hormone-releasing peptide receptor in the heart. Circ Res. 1999; 85:796-802.

15. Filigheddu N, Fubini A, Baldanzi G et al. Hexarelin protects H9c2 cardiomyocytes from doxorubicin-induced cell death. Endocrine. 2001; 14:113-9.

16. Papotti M, Ghè C, Cassoni P et al. Growth hormone secretagogue binding sites in peripheral human tissues. J Clin Endocrinol Metab. 2000; 85:3803-7.

17. Pettersson I, Muccioli G, Granata R et al. Natural (ghrelin) and synthetic (hexarelin) growth hormone secretagogues stimulate H9c2 cardiomyocyte cell proliferation. J Endocrinol. 2002; 175:201-9.

18. Kimes BW, Brandt BL. Properties of a clonal muscle cell line from rat heart. Exp Cell Res. 1976; 98:367-81.

19. Hescheler J, Meyer R, Plant S, Krautwurst D, Rosenthal W, Schultz G. Morphological, biochemical and electrophysiological characterization of a clonal cell (H9c2) from rat heart. Circ Res. 1991; 69:1476-86.

20. Chen WH, Pellegata NS, Wang PH. Coordinated effects of insulin-like growth factor-I on inhibitory pathways of cell cycle progression in cultured cardiac muscle cells. Endocrinology. 1995; 136:5240-3.

21. Ong H, Bodart V, McNicoll N, Lamontagne D, Bouchard JF. Binding sites for growth hormone-releasing peptide. Growth Hormone & IGF Res. 1998; 8:137-40.

22. Bodart V, Febbraio M, Demers A et al. CD36 mediates the cardiovascular action of growth hormone-releasing peptides in the heart. Circ Res. 2002; 90:844-9.

23. Van Nieuwenhoven FA, Luiken JJFP, De Jong YF, Grimaldi PA, Van der Vusse GJ, Glatz JFC. Stable transfection of fatty acid translocase (CD36) in a rat heart muscle cell line (H9c2). J Lipid Res. 1998; 39:2039-47.

24. Bisi G, Podio V, Valetto MR et al. Acute cardiovascular and hormonal effects of GH and hexarelin, a synthetic GH-releasing peptide, in humans. J Endocrinol Invest. 1999; 22:266-72.

25. Bisi G, Podio V, Valetto MR et al. Cardiac effects of hexarelin in hypopituitary adults. Eur J Pharmacol. 1999; 381:31-8.

26. Nagaya N, Kojima M, Uematsu M et al. Hemodynamic effects of human ghrelin in healthy volunteers. Am J Physiol Regulatory Integrative Comp Physiol. 2001; 280:R1483-7.

27. Wiley KE, Davenport AP. Comparison of vasodilators in human internal mammary artery: ghrelin is a potent physiological antagonist of endothelin-1. Brit J Pharmacol. 2002; 136:1146-52.
28. Rossoni G, Locatelli V, De Gennaro Colonna V et al. Growth hormone and hexarelin prevent endothelial vasodilator dysfunction in aortic rings of the hypophysectomized rat. J Cardiovasc Pharmacol. 1999; 34:454-60.
29. Okumura H, Nagaya N, Enomoto M, Nakagawa E, Oya H, Kangawa K. Vasodilatory effect of ghrelin, an endogenous peptide from the stomach. J Cardiovasc Pharmacol. 2002; 39:779-83.

Chapter 9

GHRELIN: CENTRAL ACTIONS AND POTENTIAL IMPLICATIONS IN NEURO-DEGENERATIVE DISEASES

Roy G. Smith, Yuxiang Sun, Alex R.T. Bailey & Antonia Paschali

Huffington Center on Aging and Departments of Molecular and Cellular Biology and Medicine, Baylor College of Medicine, Houston, Texas, Usa

Abstract: The growth hormone releasing peptides (GHRP) and their peptidomimetics were the first synthetic compounds known that stimulated both growth hormone (GH) release and appetite. Subsequently, a new orphan receptor, the growth hormone secretagogue (GHS-R), was cloned that mediated the action of these compounds. Following cloning of the GHS-R, cell lines engineered to express the cloned GHS-R provided the assays that were essential for identification of ghrelin as an endogenous GHS-R ligand in stomach tissue extracts. Ghrelin was recently also shown to be produced by a discrete network of neurons in the hypothalamus that control appetite, which are distinct from the network that regulates GH pulsatility. We describe how studies with ghrelin and its mimetics helped elucidate the mechanisms by which ghrelin regulates these distinct pathways. Localization of GHS-R expression in the brain suggests that ghrelin also modulates the function of neurons involved in memory, learning, cognition, mood and sleep. Indeed, results from studies with GHS-R (ghrelin-receptor) knockout mice indicate that reduced expression of the GHS-R causes deficits in contextual memory. This has implications for age-related neurodegeneration because ghrelin production declines in elderly adults and aging is associated with a decline in amplitude of GH-pulsatility, altered appetite, changes in metabolism and deficiencies in contextual memory. Ghrelin has the potential to prevent/reverse these changes; therefore, perhaps ghrelin is important for maintaining a young adult phenotype and providing protection against neurodegenerative diseases.

Key words: GHS-R, GH, memory, aging

1.   INTRODUCTION

The discovery of ghrelin is based on the pioneering work on synthetic growth hormone releasing peptides (GHRP) first developed in the early 1980s by Bowers and colleagues at Tulane University Medical School (1-3) and on characterization and cloning of the receptor for these small molecules by the group at Merck & Co. (4,5).

In 1988, a reverse pharmacology approach to drug discovery was initiated to identify small molecules that would increase the amplitude of growth hormone (GH) pulsatility. Although the GHRP were potent compounds, as peptides the core structure did not readily lend itself to optimization of pharmacokinetic properties. Our solution was to elucidate the GHRP-6 signal transduction pathway, identify the receptor involved, and design GHRP-6 mimetics with high potency, improved oral bioavailability and extended half-life (6-8).

In 1990, the identification of the benzolactam, L-163,429, provided the first example of the deliberate design of a nonpeptide agonist of a peptide ligand in the absence of knowledge of the receptor and its endogenous ligand (6). Like GHRP-6, L-163,429 was shown to increase the amplitude of GH-pulsatility in humans (9,10). Subsequently, an orally active mimetic, MK-0677, was developed (8). The spiropiperidine, MK-0677, was shown to be a GHRP-6 mimetic based on cross-desensitization of signal transduction and the use of selective antagonists for growth hormone releasing hormone (GHRH) and GHRP-6 (6,8). Remarkably, chronic oral administration of MK-0677 produced sustained rejuvenation of the GH/insulin-like growth factor-I (IGF-I) axis in humans (5,11,12).

The high selectivity of MK-0677 was exploited to characterize and clone a receptor for MK-0677 and the GHRP. MK-0677 was radio-labeled to high specific activity by introducing ^{35}S into the molecule and used to characterize a receptor in the anterior pituitary gland and hypothalamus (13). Taking advantage of its high affinity and high binding specificity, MK-0677 was exploited to clone the receptor. The extraordinarily low abundance of mRNA encoding the receptor necessitated development of a novel expression-cloning strategy (4). Application of this approach identified a new G-protein coupled orphan receptor that we named the growth hormone secretagogue receptor (GHS-R) (4,5).

Having cloned the GHS-R, cell lines were engineered to over-express the receptor and used to provide the essential assay for screening animal tissue extracts for a natural GHS-R ligand. Two natural agonists were identified: ghrelin in stomach extracts and adenosine in hypothalamic extracts (14-16). Like ghrelin, adenosine showed agonist activity on the GHS-R, and exhibited cross-desensitization with the synthetic ligands. However, in

contrast to ghrelin and the synthetic GHS-R agonists, adenosine failed to
stimulate GH release from pituitary cells. Hence, ghrelin, as a new hormone
and a closer mimetic of the synthetic GHS-R ligands, became the focus of
subsequent research.

2. BROAD EXPRESSION OF THE GHS-R IN THE BRAIN

Following the cloning of the porcine and human GHS-R, the rat GHS-R
was cloned and RNase protection assays were developed to investigate
GHS-R expression in different tissues (17). Transcripts were identified in the
anterior pituitary gland, hypothalamus, hippocampus and pancreas (18). In
situ hybridization with non-overlapping radiolabeled oligonucleotide probes

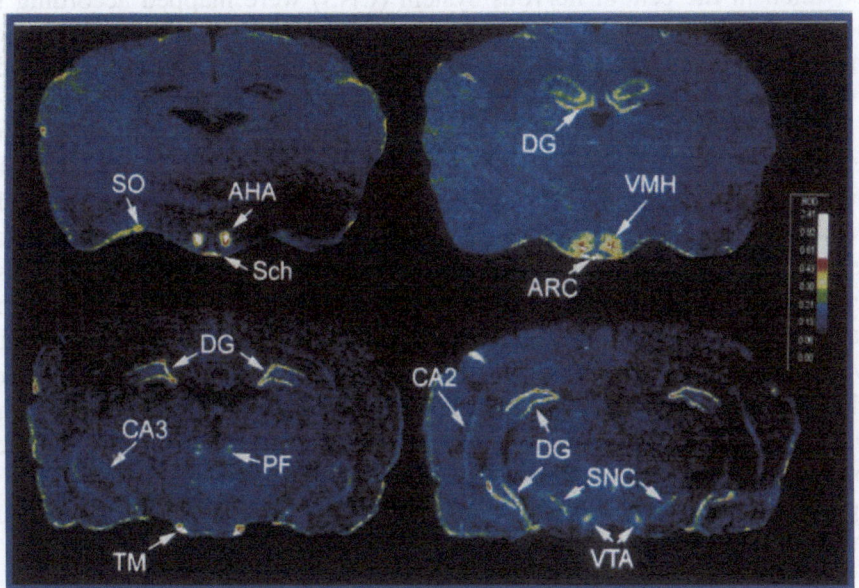

Figure 9-1. Expression of GHS-R in the brain by *in situ* hybridization (18) SO, supraoptic
nucleus; AHA, anterior hypothalamic area; Sch, suprachiasmatic nucleus; DG, dentate gyrus;
VMH, ventral medial hypothalamus; ARC, arcuate nucleus; CA3, CA2 hippocampal
structures; PF, periformical nucleus, TM, tuberomamillary nucleus; SNC, pars compacta of the
substantia nigra; VTA, ventral tegmental area. [Figure is reprinted with permission from
Elsvier. See Ref. 12]

illustrated expression of GHS-R in multiple hypothalamic nuclei including
the suprachiasmatic nucleus, anteroventral preoptic nucleus, anterior
hypothalamic area, lateroanterior hypothalamic nucleus, supraoptic nucleus,

ventromedial hypothalamic nucleus, arcuate nucleus, paraventricular nucleus and tuberomamillary nucleus (18). Expression was also observed in the dentate gyrus and CA2 and CA3 of the hippocampal formation, the pars compacta of the substantia nigra, ventral tegmental area, dorsal and medial raphae nuclei and Edinger-Westphal nucleus (18) (partially illustrated in Figure 9-1). Hence, besides neurons that play a role in the central control of GH release, the GHS-R is expressed in areas that affect appetite, biological rhythms, mood, cognition, memory, and learning (18).

3.   GHS-R IN HYPOTHALAMIC NEURONS

The role of the GHS-R in the hypothalamus was determined with synthetic ghrelin mimetics before ghrelin was identified as an endogenous ligand (3). Following administration of ghrelin mimetics, the sites of activation in the central nervous system (CNS) were mapped according to induction of the immediate-early gene c-*fos*. Peripheral administration of ghrelin mimetics in rats increased Fos expression in the hypothalamic arcuate nucleus; and electrophysiology studies showed an associated prolonged increase in the electrical activity of neurosecretory arcuate neurons (19,20). Retrograde tracing studies with FluoroGold indicated that the majority of activated cells were neurosecretory cells that project outside the blood-brain barrier (21). These include GHRH containing cells, and prior treatment with GHRH or GH markedly attenuated activation of these neurons (22,23). Additional confirmation for activation of GHRH neurons was provided by measuring concentrations of GHRH and somatostatin in hypothalamic/pituitary portal vessels during administration of ghrelin mimetics to sheep; the concentrations of GHRH increased, whereas somatostatin levels did not change markedly (24-26). However, this does not apply to all species. For example, when similar studies were performed in pigs, GHRH levels were unchanged in the portal vessels, but somatostatin levels were decreased in response to the ghrelin mimetic (27).

Ghrelin itself has also been shown to activate early-response genes (egr and Fos) in the arcuate nucleus following systemic administration (29). Similar studies using double labeling immunohistochemistry showed that Fos was activated in GHRH and neuropeptide Y (NPY) arcuate neurons, which was identical to that observed previously with ghrelin mimetics (28,29).

We were interested in learning whether a truncated form of ghrelin, which could be more readily modeled on the ghrelin mimetic structures, would be a potent activator of arcuate neurons. Therefore, we compared the *in vivo* potencies of MK-0677, ghrelin and ghrelin 1-14 when administered intraperitoneally to mice (Figure 9-2). Ghrelin was found to have similar potency to MK-0677 for activating Fos in arcuate neurons, whereas ghrelin 1-14 was less active. The lower potency of ghrelin 1-14 in the circulation would be explained if the octanoyl ester linkage in the truncated peptide is more susceptible to hydrolysis by esterases. However, although this result suggests that full-length ghrelin is the biologically important peptide, our experiments do not preclude the truncated forms having biological significance when produced locally.

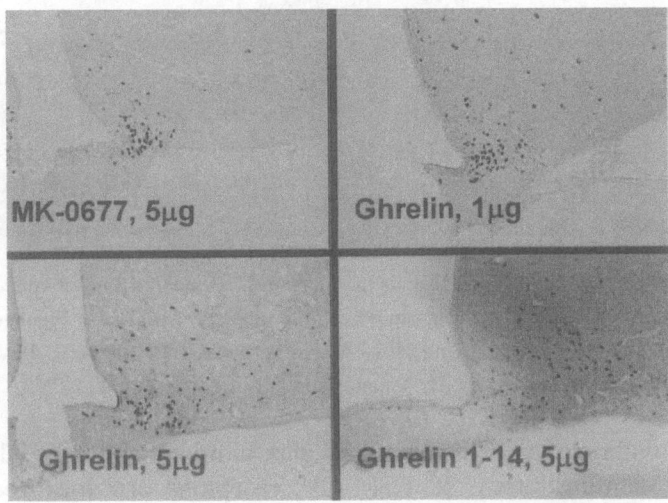

Figure 9-2. Activation of Fos in the arcuate nucleus following intraperitoneal injection of mice with MK-0677, ghrelin and ghrelin 1-14 [Bailey ART and Smith RG, unpublished work using protocol described previously in mice (22)].

3.1   Ghrelin regulation of pulsatile GH-release

Probably the best characterized function of ghrelin and its mimetics is its property of increasing the amplitude of pulsatile GH-release. The main locus of activity is stimulation of GHRH release from neurons in the arcuate nucleus (Figure 9-3); however, the ghrelin mimetics also increase GH release directly by amplifying the effects of GHRH and antagonizing somatostatin on the anterior pituitary gland (7). NPY neurons in the arcuate nucleus are also activated by ghrelin and its mimetics (30,31); the majority of these NPY neurons project outside the blood-brain barrier and release NPY into the hypothalamic-pituitary portal vessels (21,32). Remarkably,

NPY acts on somatotrophs to increase the release of GH (33), which provides an additional indirect amplification of GH-release by ghrelin receptor agonists. Hence, NPY has the properties of the putative hypothalamic "U-factor" proposed by Bowers (34).

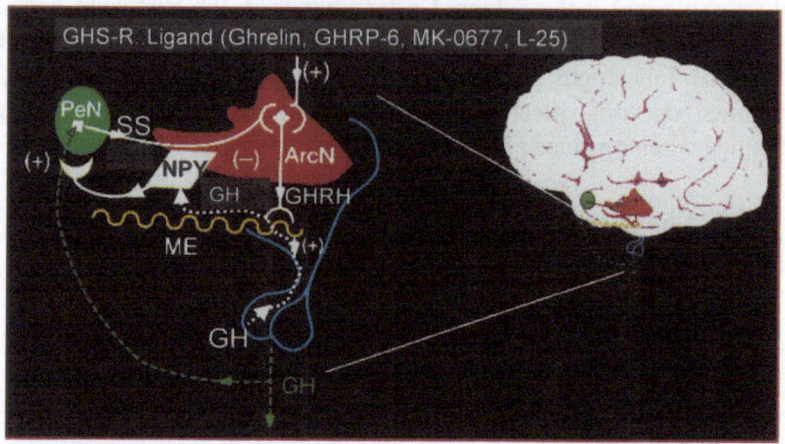

Figure 9-3. Model of how GHS-R ligands modulate the GH mediated negative feedback loop by their action on arcuate neurons and how the pathway might be regulated by both somatostatin and NPY. Arcuate nucleus, ArcN; periventricular nucleus, PeVN; median eminence, ME; growth hormone, GH; somatostatin, SS; neuropeptide Y, NPY.

The activation of arcuate neurons by ghrelin mimetics can be inhibited by peripheral administration of GH. GH receptors are localized in the periventricular nucleus (PeVN) and arcuate nucleus (35)-38). Microinjection of GH into the PeVN or into the arcuate nucleus reduces endogenous GH-release (36). In hypophysectomized rats, GH treatment induces c-*fos* expression in somatostatin neurons in the PeVN and NPY neurons in the arcuate nucleus, which suggests that both these neuropeptides are involved in regulating activity of arcuate neurons that express ghrelin receptors (37-39). To evaluate the role of somatostatin and to determine which receptor subtype was involved in the negative feedback, Zheng and colleagues generated somatostatin receptor subtype-2 (sst2) null mice (22). Wild-type and sst2 null mice were injected with mouse GH or vehicle before injection with a ghrelin mimetic. Activation of Fos in the PeVN was observed with both genotypes; however, GH-pretreatment inhibited activation of Fos in arcuate neurons of wild-type mice but not in sst2-null mice. These results are consistent with GH-regulated negative feedback being controlled by activation of the sst2 by GH-induced somatostatin release from neurons in the PeVN (22). In addition, there is compelling evidence that GH activation of NPY neurons in the arcuate nucleus is implicated in the negative feedback pathway (40,41). Hence, a dual pathway appears to be involved where arcuate GHRH neurons are inhibited either by somatostatin and NPY independently, or by NPY acting on somatostatin neurons to maintain stimulation of somatostatin tone.

3.2   Activity in Paraventricular Nucleus and Supraoptic Nucleus

As well as inducing Fos in GHRH neurons, ghrelin and ghrelin mimetics increase *c-fos* mRNA in NPY neurons (20,31). To evaluate the consequences of activating these arcuate neurons, it is important to determine where the NPY neurons project. By retrograde tracing and *in situ* hybridization studies, it was established that a subset of arcuate NPY neurons project to the parvocellular region of the ipsilateral paraventricular nucleus (PVN) (42). Using a combination of electrophysiology and retrograde tracing, Honda and colleagues investigated activation of neurosecretory arcuate neurons that project to the PVN and median eminence (43). They tested 116 arcuate neurons and 43 of them were identified antidromically as projecting to the PVN, and 30 projected to the median eminence. Each population displayed distinct electrophysiological properties and patterns of orthodromic response following stimulation of the median eminence and PVN; hence, the two populations appeared functionally distinct. About one-third of the arcuate neurons that projected to the PVN were found to be activated by systemic injection of the ghrelin mimetic GHRP-6. In a parallel series of experiments, FluoroGold retrograde-labeling showed that Fos was induced in 20% of the arcuate neurons that project to the PVN. Although these results confirmed the existence of direct projections from the arcuate nucleus to the PVN and showed that the ghrelin mimetic activated some of these neurons, less than 5% of the total number of Fos-positive arcuate neurons were both Fos-positive and retrogradely-labeled.

Acute administration of ghrelin and ghrelin mimetics stimulate the release of vasopressin from hypothalamic neurons, which potentially explains the observed increases in serum glucocorticoid levels through activation of the hypothalamic-pituitary-adrenal (HPA) axis (44,45). The cell bodies of vasopressin neurons are located in the PVN and supraoptic nucleus (SON), but Fos induction in these nuclei associated with ghrelin treatment has never been reported.

Intriguingly, although ghrelin, MK-0677 and GHRP-6 do not readily increase Fos in the PVN or the SON of rats or mice, the structurally distinct highly potent ghrelin mimetic, GHS-25, produced a robust Fos response in the SON of rats (Figure 9-4). Double-labeling immunohistochemistry revealed that approximately 66% of Fos-labeled neurons contained oxytocin, 15% contained vasopressin, and the remaining cells contained neither oxytocin nor vasopressin (23). These results indicate that besides regulating GHRH and NPY release, ghrelin has the capacity to control neurohypophyseal hormone release.

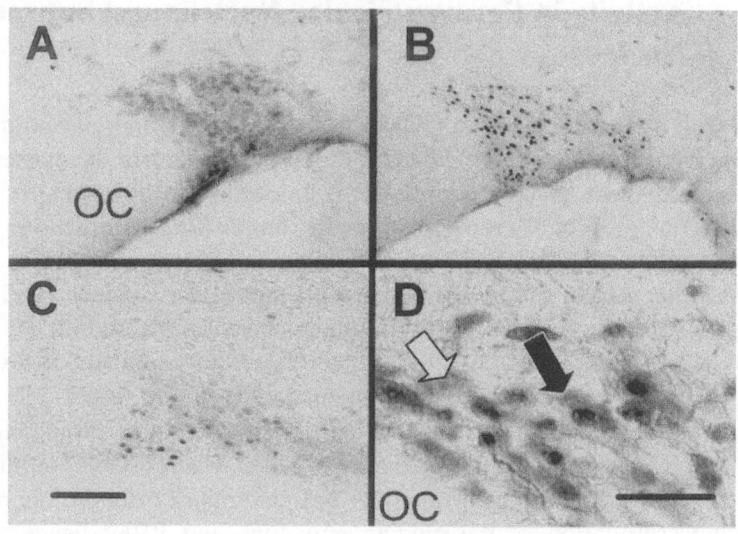

Figure 9-4. Photomicrographs of the rat supraoptic nuclei (SON) of male rats showing Fos immunoactivity following administration of (A) GHRP-6 (50 µg); (B) the ghrelin mimetic, GHS-25 (50 µg); (C) GHRH (2 µg)/GHS-25 (50 µg); scale bar = 0.2 mm. (D) Photomicrograph of the SON showing immunocytochemical double labeling for Fos and oxytocin; the white arrow points to a cell with cytoplasmic labeling for oxytocin alone, and the black arrow points to a doubly-labeled cell with cytoplasm labeled for oxytocin and nucleus labeled for Fos; scale bar = 100 µm. OC, optic chiasm (23). (Figure is reprinted with permission from Humana Press).

Curiously, in contrast to the rat studies, systemic administration of GHS-25 to mice increased Fos immunoactivity in the PVN, but not the SON (Figure 9-5, Bailey and Smith, unpublished work). These data illustrate the need for caution in making general conclusions from data derived from different species. Although activation of the SON and PVN was not detected following the systemic administration of ghrelin to either rats or mice, activation of these nuclei by GHS-25 is still likely to be relevant for ghrelin action. GHS-25 is a highly specific agonist for the GHS-R, and extensive site-directed mutagenesis studies on the GHS-R comparing the activity of ghrelin and GHS-25 on different GHS-R mutants show that GHS-25 is a closer mimetic of ghrelin than either GHRP-6 or MK-0677 (Smith et al., unpublished results); therefore, the inability to visualize activation of Fos in the SON and PVN by ghrelin might be a reflection of increased *in vivo* potency of GHS-25 and/or the different physical properties of GHS-25, such as its higher stability and improved penetration of the blood-brain barrier.

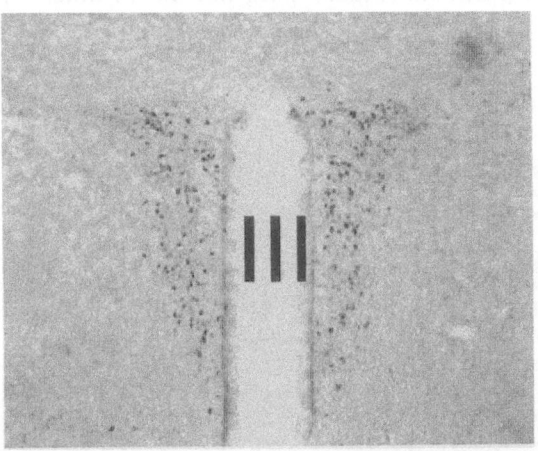

Figure 9-5 Photomicrographs of the mouse paraventricular nucleus (PVN) showing Fos immunoactivity following administration of the ghrelin mimetic, GHS-25 (5 µg); III, third ventrical (Bailey and Smith, unpublished results).

4.   ACTIVATION OF BRAINSTEM NUCLEI AND IMPLICATION OF THE VAGUS NERVE IN SIGNALING

The arcuate nucleus is believed to reside within the blood-brain barrier (46); therefore, a question frequently raised is, how do ghrelin and ghrelin mimetics gain access to the CNS and activate arcuate neurons? One possibility is that information is relayed through the circumventricular organs of the midbrain (47,48); however, activation of Fos in this area of the brain has not been observed following the administration of ghrelin mimetics (49,50). Alternatively, signaling could be mediated via the brainstem; for example, the area postrema, which is connected to the nucleus tractus solitarii (NTS), includes noradrenergic neurons that project to the arcuate nucleus and supraoptic nucleus (51,52). Furthermore, these projections appear to regulate the release of hypothalamic peptides including oxytocin, vasopressin and GHRH (52-54).

Based on the above considerations, Bailey and colleagues systematically investigated whether ghrelin mimetics activate Fos in the brainstem (55). They observed induction of Fos expression following peripheral injection of the ghrelin mimetics, GHRP-6 and MK-0677. The Fos-positive nuclei were detected in the area postrema (Figure 9-6), but not in other regions of the brainstem.

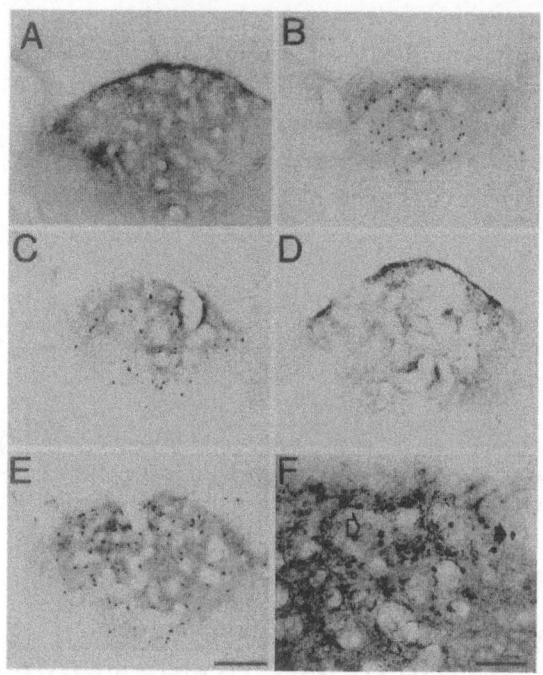

Figure 9-6. Photomicrographs illustrating Fos protein expression in the area postrema of male rats injected with (A) GHRH (2 μg); (B) GHRP-6 (50 μg); (C) MK-0677 (50 μg); (D) sandostatin (100 μg); (E) sandostatin (100 μg)/MK-0677 (50 μg); scale bar = 0.2 mm. (F) Photomicrograph showing double labeling for Fos and tyrosine hydroxylase in the area postrema of a rat injected with GHRP-6 (50 μg). The filled arrow points to a cell stained only for Fos and the open arrow points to a cell stained only for tyrosine hydroxylase; scale bar = 500 μm (55). (Figure is reprinted with permission from Blackwell Publishing).

To determine whether Fos induction might be a result of the stimulation of endogenous GHRH or GH release by ghrelin mimetics, rats were injected with a dose of GHRH known to maximally stimulate GH release. GHRH treatment failed to induce Fos expression in the area postrema, NTS or other areas of the brainstem. Prior injection of the somatostatin agonist sandostatin neither induced Fos expression in the brainstem nor blocked ghrelin mimetic induced Fos expression in the area postrema (Figure 9-6). These results indicate that activation of the area postrema is not secondary to activation of

GHRH arcuate neurons or hypothalamic somatostatin neurons and is, therefore, distinct from the GH-releasing properties of the ghrelin mimetics.

To establish whether signaling between the brainstem and arcuate neurons was mediated by a noradrenergic pathway, the induction of Fos expression by ghrelin mimetics was compared in intact and noradrenergic-lesioned rats (55). The rats were lesioned by injection of a specific neurotoxin, 5-amino-2,4-dihydroxy-α-methylphenylethylamine (ADMP), into the right lateral ventricle. After five days, the lesioned and mock-lesioned (aCSF vehicle) rats were treated systemically with a ghrelin mimetic, and their brains harvested and sectioned. Alternate sections were evaluated by immunocytochemistry for either Fos or dopamine-β-hydroxylase (DBH). DBH-positive stained fibers were observed throughout the hypothalamus of the control rats with marked reductions evident in the ADMP treated animals; however, the level of Fos induced in the arcuate nucleus by treatment with the ghrelin mimetic was identical in ADMP-treated and control rats. Hence, these adrenergic pathways are apparently not involved in ghrelin mediated activation of arcuate neurons.

Perhaps activation of the brainstem is not involved in the GH-releasing properties of ghrelin. Besides GH-release, acute treatment of animals with ghrelin or ghrelin mimetics increases appetite (56-59) and activates the HPA axis, resulting in increased adrenocorticotrophic hormone (ACTH) and glucocorticoid release (60). The area postrema and NTS are involved in gastric signaling (61,62) and projections from the NTS to the paraventricular nucleus are important for activation of the HPA axis (63,64).

Ghrelin is primarily produced in the stomach and sensory information can be transmitted to the brain through the afferent vagus nerve; therefore, the potential role of the gastric afferent vagal nerve in ghrelin signaling was evaluated (29). The ghrelin receptor was localized in vagal afferent neurons by reverse-transcription polymerase chain reaction and *in situ* hybridization. It appeared that the receptor is synthesized in vagal neurons and transported to the afferent terminals. When ghrelin was administered systemically to vagotomized rats, or following perivagal application of the afferent neurotoxin capsaicin, ghrelin failed to stimulate appetite GH secretion and activation of Fos in hypothalamic NPY and GHRH neurons. It was concluded that the gastric vagal afferent is the major pathway conveying ghrelin's signals to the brain (29). Intracerebral ventricular administration of ghrelin also induces Fos expression in the NTS and dorsomotor nucleus of the vagus, which are important sites in the CNS for regulation of gastric acid secretion (65). These results implicate the vagus nerve and ghrelin in central regulation of appetite and gastric acid secretion.

5. GHRELIN AND ENERGY HOMEOSTASIS

The localization and production of ghrelin in the stomach focused attention on ghrelin's potential role in obesity. Plasma ghrelin levels are influenced by nutritional status and are believed to regulate GH, appetite, and fat deposition (59,66-69). Most intriguing was the observation that low circulating ghrelin levels correlate with sustained weight loss and reduced appetite in obese humans following gastric bypass surgery (70). However, whether these beneficial changes are a result of reduced ghrelin, rather than altered production of other gut peptides that regulate appetite, is unclear. An association with obesity was surprising because chronic administration of MK-0677 to obese humans caused increases in lean mass, but had no effect on fat (71).

Leptin and ghrelin have opposite effects on appetite, and it has been shown that ghrelin-excited neurons in the ventromedial arcuate nucleus (VMH) are inhibited by leptin (72). The hypothalamic leptin- and ghrelin-regulated network appears to involve NPY and melanocyte stimulating hormone (MSH). NPY neurons are located mainly in the VMH, and the pro-opiomelanocortin (POMC) neurons, which produce MSH, predominate in the ventrolateral region. Riediger and colleagues showed, using extracellular recordings from arcuate nucleus slice preparations, that ghrelin (10^{-8} M) was excitatory for the majority of VMH neurons, whereas, 42% of the ventrolateral arcuate neurons were inhibited (73). Ghrelin's excitatory effect appears to be post-synaptic because it was not blocked by synaptic blockade, whereas, ghrelin's inhibitory effect was prevented by blocking synaptic interactions. These results strongly suggest that ghrelin opposes leptin action by directly stimulating NPY neurons in the VMH and indirectly inhibiting POMC neurons in the ventral lateral nucleus (73).

A recent study addressed the question of whether ghrelin was expressed in areas of the hypothalamus that are involved in regulating energy balance (74). Ghrelin-immunoreactive cells were identified that fill the internuclear space between the lateral (LH) arcuate, VMH, dorsomedial (DMH), PVN, and the ependymal layer of the third ventricle. This unique distribution does not overlap with known hypothalamic cell populations, such as those that produce NPY, agouti-related protein (AgRP), POMC, melanin concentrating hormone, orexin, dopamine, and somatostatin 8-14. These observations suggest specific roles for locally produced ghrelin in the CNS.

Immuno-electron microscopy showed that ghrelin is located in axons where it is associated with dense-cored vesicles in presynaptic terminals (74). These axon terminals innervate the arcuate nucleus, DMH, LH, PVN and ghrelin boutons, and appear to make synaptic contact with cell bodies, dendrites of NPY/AgRP and POMC neurons in the arcuate nucleus, and

NPY and GABA axon terminals in the arcuate nucleus and PVN. Such interactions suggest a presynaptic mode of action for ghrelin in the hypothalamus. Some ghrelin axons in the PVN innervate corticotropin-releasing hormone (CRH) cells, which is consistent with the increase in ACTH and glucocorticoid secretion observed following treatment with ghrelin and its mimetics. These observations delineate an anatomical basis for pre- and post-synaptic interactions between ghrelin and NPY/AgRP, POMC, and CRH circuits (Figure 9-7).

Hypothalamic localization of the GHS-R was investigated in coronal slices of rat brain using biotin-labeled ghrelin (74). Binding of biotinylated ghrelin was observed in the arcuate nucleus, LH, and PVN, and was mainly associated with pre-synaptic boutons. Axon terminals that bound ghrelin were frequently found to contain NPY. Together, the binding data and the localization of expression of ghrelin in axons adjacent to pre-synaptic nerve terminals support the notion that ghrelin modulates neurotransmission.

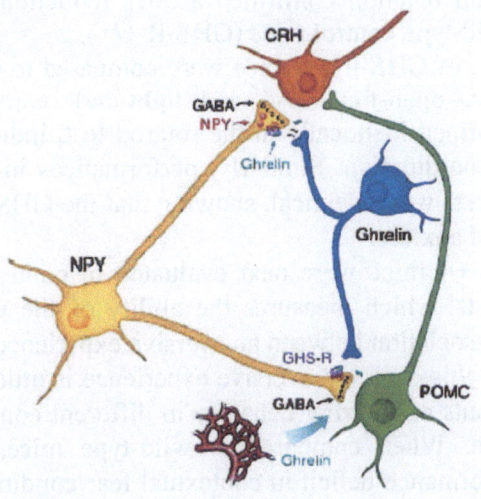

Figure 9-7 Model of the interaction between ghrelin and hypothalamic circuits (74). (Reprinted with permission from Elsevier)

In summary, ghrelin is produced in the hypothalamus where it is localized to a previously uncharacterized group of neurons adjacent to the third ventricle between the dorsal, ventral, paraventricular and arcuate hypothalamic nuclei (74). These neurons send efferents onto key hypothalamic circuits, which include those producing NPY, AgRP, POMC products, and CRH. In the hypothalamus, ghrelin binds mainly to pre-synaptic terminals of NPY neurons. Electrophysiological recordings showed that ghrelin stimulated the activity of arcuate NPY neurons, and mimicked the effect of NPY in the PVN. The authors propose that at these sites, release of ghrelin stimulates the release of orexigenic peptides and neurotransmitters, thus representing a novel regulatory circuit controlling energy homeostasis (Figure 9-7).

6.   GHRELIN AND GHS-R IN THE HIPPOCAMPUS

Expression of the GHS-R has been demonstrated in the dentate gyrus, and CA2 and CA3 hippocampal structures. To address whether the GHS-R might be important for hippocampal function, we generated GHS-R knockout mice. To minimize differences in development that might occur as a consequence of the complete absence of the GHS-R, we selected GHS-R heterozygotes (GHS-R +/-), rather than homozygotes, for behavioral studies. Expression of GHS-R by real-time quantitative reverse transcriptase-polymerase chain reaction confirmed a 50% reduction in GHS-R mRNA compared to wild-type control mice (GHS-R +/+).

The behavior of GHS-R +/- mice were compared to GHS-R +/+ mice in the rotarod test, open-field test, and light-dark exploration (75). Both genotypes performed identically in the rotarod test, indicating the mice had normal motor coordination. Similarly, performances in the open field and light/dark box tests were identical, showing that the GHS-R +/- mice did not exhibit increased anxiety.

The GHS-R +/- mice were next evaluated in contextual and cued fear conditioning tests, which measures the ability of the mouse to learn and remember the association between an aversive experience and environmental cues (76-78). In this case, the aversive experience is mild foot-shock; fear is monitored as bouts of freezing behavior in different contexts and following an auditory cue. When compared to wild-type mice, GHS-R +/- mice exhibited a performance deficit in contextual fear conditioning, but behaved identically in the cued fear conditioning test (Figure 9-7). Hence, the GHS-R +/- mice are able to associate an auditory cue with a foot-shock, but cannot associate where that shock took place, which illustrates that they can learn simple but not more complicated cues.

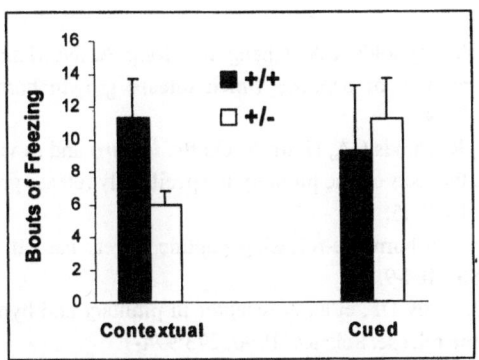

Figure 9-8. Contextual and Cued Fear Conditioning in GHS-R +/+ and GHS-R +/- Mice. Error bars show standard error of the mean.

The deficit in contextual fear conditioning in the GHS-R +/- mice is consistent with functional differences in the hippocampal formation (79). Impaired function of the GHS-R +/- mice was also indicated in the Morris water maze; however, these changes failed to reach statistical significance. These results are consistent with an important role for the GHS-R in hippocampal function.

7. SUMMARY

The discovery of ghrelin was dependent upon the development of the GHRP and cloning of the orphan GHS-R. Like the GHRP, ghrelin amplifies the amplitude of GH-release and appetite and the activity of hypothalamic neurons that regulate these pathways. Moreover, the pathways activated by ghrelin are associated directly and indirectly with signaling to peripheral tissues and centers of the brain most affected by aging and age-related neurodegeneration. Recent studies in humans show that ghrelin levels decline during aging (80). The demonstration that the GH/IGF-I axis can be rejuvenated by chronic once-daily oral treatment with a ghrelin mimetic such as MK-0677, and that reduced expression of the GHS-R results in deficits in contextual memory, support the concept that maintaining ghrelin tone may have beneficial effects on the aging process, particularly in the CNS.

REFERENCES

1. Momany FA, Bowers CY, Reynolds GA, Chang D, Hong A, Newlander K. Design, synthesis, and biological activity of peptides which release growth hormone, *in vitro*. Endocrinology. 1981; 108:31-9.
2. Bowers CY, Momany FA, Reynolds GA, Hong A. On the *in vitro* and *in vivo* activity of a new synthetic hexapeptide that acts on the pituitary to specifically release growth hormone. Endocrinology. 1984; 114:1537-45.
3. Bowers CY Unnatural growth hormone-releasing peptide begets natural ghrelin. J Clin Endocrinol Metab. 2001; 86:1464-9.
4. Howard AD, Feighner SD, Cully DF, et al. A receptor in pituitary and hypothalamus that functions in growth hormone release. Science. 1996; 273:974-7.
5. Smith RG, Van der Ploeg LH, Howard AD, et al. Peptidomimetic regulation of growth hormone secretion. Endocr Rev. 1997; 18:621-45.
6. Smith RG, Cheng K, Schoen WR, et al. A nonpeptidyl growth hormone secretagogue. Science. 1993; 260:1640-3.
7. Smith RG, Pong S-S, Hickey GJ, et al. Modulation of pulsatile GH release through a novel receptor in hypothalamus and pituitary gland. Rec Prog Horm Res. 1996; 51:261-86.
8. Patchett AA, Nargund RP, Tata JR, et al. The design and biological activities of L-163,191 (MK-0677): a potent orally active growth hormone secretagogue. Proc Natl Acad Sci USA. 1995; 92:7001-5.
9. Huhn WC, Hartman ML, Pezzoli SS, Thorner MO. 24-hour growth hormone (GH)-releasing peptide (GHRP) infusion enhances pulsatile GH secretion and specifically attenuates the response to a subsequent GHRP-6 bolus. J Clin Endocrinol Metab. 1993; 76:1202-08.
10. Chapman IM, Hartman ML, Pezzoli SS, Thorner MO. Enhancement of pulsatile growth hormone secretion by continuous infusion of a growth hormone-releasing peptide mimetic, L-692,429, in older adults - a clinical research center study. J Clin Endocrinol Metab. 1996; 81:2874-80.
11. Chapman IM, Bach MA, Van Cauter E, et al. Stimulation of the growth hormone (GH)-insulin-like growth factor-I axis by daily oral administration of a GH secretagogue (MK-0677) in healthy elderly subjects. J Clin Endocrinol Metab. 1996; 81:4249-57.
12. Smith RG, Feighner S, Prendergast K, Guan X, Howard A. A new orphan receptor involved in pulsatile growth hormone release. Trends Endocrinol Metab. 1999; 10:128-35.
13. Pong S-S, Chaung L-Y, Dean DC, Nargund RP, Patchett AA, Smith RG. Identification of a new G-protein-linked receptor for growth hormone secretagogues. Mol Endocrinol. 1996; 10:57-61.
14. Kojima M, Hosoda H, Date Y, Nakazato M, Matsuo H, Kangawa K. Ghrelin is a growth-hormone-releasing acylated peptide from stomach. Nature. 1999; 402:656-60.
15. Tullin S, Hansen BS, Ankersen M, Moller J, Von Cappelen KA, Thim L. Adenosine is an agonist of the growth hormone secretagogue receptor. Endocrinology. 2000; 141:3397-402.
16. Smith RG, Griffin PR, Xu Y, et al. Adenosine: A partial agonist of the growth hormone secretagogue receptor. Biochem Biophys Res Commun. 2000; 276:1306-13.

17. McKee KK, Palyha OC, Feighner SD, et al. Molecular analysis of growth hormone secretagogue receptors (GHS-Rs): cloning of rat pituitary and hypothalamic GHS-R type 1a cDNAs. Mol Endocrinol. 1997; 11:415-23.
18. Guan XM, Yu H, Palyha OC, et al. Distribution of mRNA encoding the growth hormone secretagogue receptor in brain and peripheral tissues. Brain Res Mol Brain Res. 1997; 48:23-9.
19. Dickson SL, Leng G, Robinson ICAF. Systemic administration of growth hormone-releasing peptide (GHRP-6) activates hypothalamic arcuate neurones. Neuroscience. 1993; 53:303-6.
20. Dickson SL, Luckman SD. Induction of c-fos messenger ribonucleic acid in neuropeptide-Y and growth hormone (GH)-releasing factor neurones in the rat arcuate nucleus following systemic injection of the GH secretagogue, GH-releasing peptid-6. Endocrinology. 1997; 138:771-7.
21. Dickson SL, Doutrelant-Viltart O, Dyball REJ, Leng G. Retrogradely labelled neurosecretory neurones of the rat hypothalamic arcuate nucleus express Fos protein following systemic injection of growth hormone (GH)-releasing peptide. J Endocr. 1996; 151:323-31.
22. Zheng H, Bailey ART, Jiang M-H, et al. Somatostatin receptor subtype-2 knockout mice are refractory to growth hormone negative feedback on arcuate neurons. Mol Endocrinol. 1997; 11:1709-17.
23. Bailey ART, Gilliver L, Leng G, Smith RG. Central actions of a nonpeptide growth hormone secretagogue GHS-25. Endocrine. 2001; 14:15-9.
24. Guillaume V, Magnan E, Cataldi M, et al. Growth hormone (GH)-releasing hormone secretion is stimulated by a new GH-releasing hexapeptide in sheep. Endocrinology. 1994; 135:1073-6.
25. Fletcher TP, Thomas GB, Clarke IJ. Growth hormone-releasing hormone and somastatin concentrations in the hypophysial portal blood of conscious sheep during the infusion of growth hormone-releasing peptide-6. Domestic Animal Endocrinol. 1996; 13:251-8.
26. Smith RG, Cheng K, Pong S-S, et al. Mechanism of action of GHRP-6 and nonpeptidyl growth hormone secretagogues. In: Bercu BB, Walker RF, eds. Growth Hormone Secretagogues, Serono Symposia. New York: Springer-Verlag. 1996; 147-63.
27. Drisko JE, Faidley TD, Zhang D, et al. Administration of a nonpeptidyl growth hormone secretagogue, L-163, 255, changes somatostatin pattern, but has no effect on patterns of growth hormone-releasing factor in the hypophyseal-portal circulation of the conscious pig. Proc Soc Exp Biol Med. 1999; 222:70-7.
28. Hewson AK, Dickson SL. Systemic administration of ghrelin induces Fos and Egr-1 proteins in the hypothalamic arcuate nucleus of fasted and fed rats. J Neuroendocrinol. 2000; 12:1047-9.
29. Date Y, Murakami N, Toshinai K, et al. The role of the gastric afferent vagal nerve in ghrelin-induced feeding and growth hormone secretion in rats. Gastroenterology. 2002; 123:1120-8.
30. Willesen MG, Kristensen P, Romer J. Co-localization of growth hormone secretagogue receptor and NPY mRNA in the arcuate nucleus of the rat. Neuroendocrinology. 1999; 70:306-16.

31. Wang L, Saint-Pierre DH, Tache Y. Peripheral ghrelin selectively increases Fos expression in neuropeptide Y - synthesizing neurons in mouse hypothalamic arcuate nucleus. Neurosci Lett. 2002; 325:47-51.

32. McDonald JK, Koenig JI, Gibbs DM, Collins P, Noe BD. High concentrations of neuropeptide Y in pituitary portal blood of rats. Neuroendocrinology. 1987; 46:538-41.

33. McDonald JK, Lumpkin MD, Samson WK, McCann SM. Neuropeptide Y affects secretion of luteinizing hormone and growth hormone in ovariectomized rats. Proc Natl Acad Sci USA. 1985; 82:561-4.

34. Bowers CY. Xenobiotic growth hormone secretagogues: growth hormone releasing peptides. In: Bercu BB, Walker RF, eds. Growth Hormone Secretagogues, Serono Symposia. New york: Springer-Verlag. 1996; 9-28.

35. Pelligrini E, Bluet-Pajot MT, Mounier F, Bennet P, Cordon C, Epelbaum J. Central administraiton of growth hormone receptor mRNA antisense increases GH pulsatility and decreases hypothalamic somatostatin expression in rats. J Neurosci. 1996; 16:8140-8.

36. Minami S, Suzuki N, Sugihara H, Tamura H, Emoto N, Wakabashi I. Microinjection of rat GH but not human IGF-I into a defined area of the hypothalamus inhibits endogenous GH secretion in rats. Journal of Endocrinology. 1997; 153:283-90.

37. Minami S, Kamegai J, Sugihara H, Hasegawa O, Wakabayashi I. Systemic administration of recombinant human growth hormone induces expression of the c-fos gene in the hypothalamic arcuate and periventricular nuclei in hypophysectomized rats. Endocrinology. 1992; 131:247-53.

38. Kamegai J, Minami S, Sugihara H, Higuchi H, Wakabayashi I. Growth hormone induces expression of the c-fos gene on hypothalamic neuropeptide-Y and somatostatin neurons in hypophysectomized rats. Endocrinology. 1994; 6:2765-71.

39. Kamegai J, Minami S, Sugihara H, Hasegawa O, Higuchi H, Wakabayashi I. Growth hormone receptor gene is expressed in neuropeptide Y neurons in hypothalamic arcuate nucleus of rats. Endocrinology. 1996; 137:2109-12.

40. Chan YY, Steiner RA, Clifton DK. Regulation of hypothalamic neuropeptide-Y neurons by growth hormone in the rat. Endocrinology. 1996; 137:1319-25.

41. Chan YY, Clifton DK, Steiner RA. Role of NPY neurones in GH-dependent feedback signalling to the brain. Horm Res. 1996; 45:12-4.

42. Baker RA, Herkenham M. Arcuate nucleus neurons that project to the hypothalamic paraventricular nucleus: neuropeptidergic identity and consequences of adrenalectomy on mRNA levels in the rat. J Comp Neurol. 1995; 358:518-30.

43. Honda K, Bailey ART, Bull PM, MacDonald LP, Dickson SL, Leng G. An electrophysiological and morphological investigation of the projections of growth hormone-releasing peptide-6-responsive neurons in the rat arcuate nucleus to the median eminence and to the paraventricular nucleus. Neuroscience. 1999; 90:875-83.

44. Korbonits M, Kaltsas G, Perry LA, et al. The growth hormone secretagogue hexarelin stimulates the hypothalamo-pituitary-adrenal axis via arginine vasopressin. J Clin Endocrinol Metab 1999; 84:2489-95.

45. Korbonits M, Little JA, Forsling ML, et al. The effect of growth hormone secretagogues and neuropeptide Y on hypothalamic hormone release from acute rat hypothalamic explants. J Neuroendocrinol. 1999; 11:521-8.

46. Merchenthaler I. Neurons with access to the general circulation in the central nervous system of the rat: a retrograde tracing study with fluoro-gold. Neuroscience. 1991; 44:655-62.

47. Ferguson AV, Bains JS. Electrophysiology of the circumventricular organs. Front Neuroendocrinol. 1996; 17:440-75.

48. Johnson AK, Gross PM. Sensory circumventricular organs and brain homeostatic pathways. Faseb J. 1993; 7:678-86.

49. Bailey ART, Smith RG, Leng G. The non-peptide growth hormone secretagogue, MK-0677, activates hypothalamic neurones *in vivo*. J Neuroendocrinol. 1998; 10:111-8.

50. Dickson SL, Leng G, Dyball REJ, Smith RG. Central actions of peptide and nonpeptide growth hormone secretagogues in the rat. Neuroendocrinology. 1995; 61:36-43.

51. Sawchenko PE, Swanson LW. Central noradrenergic pathways for the integration of hypothalamic neuroendocrine and autonomic responses. Science. 1981; 214:685-7.

52. Day TA, Sibbald JR. Direct catecholaminergic projection from nucleus tractus solitarii to supraoptic nucleus. Brain Res. 1988; 454:387-92.

53. Onaka T, Luckman SM, Antonijevic I, Palmer JR, Leng G. Involvement of the noradrenergic afferents from the nucleus tractus solitarii to the supraoptic nucleus in oxytocin release after peripheral cholecystokinin octapeptide in the rat. Neuroscience. 1995; 66:403-12.

54. Makara GB, Kiem DT, Vizi ES. Hypothalamic alpha 2A-adrenoceptors stimulate growth hormone release in the rat. Eur J Pharmacol. 1995; 287:43-8.

55. Bailey ART, Von Englehardt N, Smith RG, Leng G, Dickson SL. Growth hormone secretagogue activation of the arcuate nucleus and brainstem occurs via a non-noradrenergic pathway. J Neuroendocrinol. 2000; 12:191-8.

56. Locke W, Kirgis HD, Bowers CY, Abdoh AA. Intracerebroventricular growth hormone-releasing peptide-6 stimulates eating without affecting plasma growth hormone responses in rats. Life Sci. 1995; 56:1347-52.

57. Okada K, Ishii S, Minami S, Sugihara H, Shibasaki T, Wakabayashi I. Intracerebroventricular administration of the growth hormone-releasing peptide KP-102 increases food intake in free-feeding rats. Endocrinology. 1996; 137:5155-8.

58. Wren AM, Small CJ, Ward HL, et al. The novel hypothalamic peptide ghrelin stimulates food intake and growth hormone secretion. Endocrinology. 2000; 141:4325-8.

59. Nakazato M, Murakami N, Date Y, et al. A role for ghrelin in the central regulation of feeding. Nature. 2001; 409:194-8.

60. Frieboes RM, Murck H, Schier T, Holsboer F, Steiger A. Somatostatin impairs sleep in elderly human subjects. Neuropsychopharmacology. 1997; 16:339-45.

61. Miller AD, Leslie RA. The area postrema and vomiting. Front Neuroendocrinol. 1994; 15:301-20.

62. Dockray GJ. The G. W. Harris Prize Lecture. The gut endocrine system and its control. Exp Physiol. 1994; 79:607-34.

63. Herman JP, Cullinan WE. Neurocircuitry of stress: central control of the hypothalamo-pituitary-adrenocortical axis. Trends Neurosci. 1997; 20:78-84.

64. Herman JP, Prewitt CM, Cullinan WE. Neuronal circuit regulation of the hypothalamo-pituitary-adrenocortical stress axis. Crit Rev Neurobiol. 1996; 10:371-94.

65. Date Y, Nakazato M, Murakami N, Kojima M, Kangawa K, Matsukura S. Ghrelin acts in the central nervous system to stimulate gastric acid secretion. Biochem Biophys Res Commun. 2001; 280:904-7.

66. Peino R, Baldelli R, Rodriguez-Garcia J, et al. Ghrelin-induced growth hormone secretion in humans. Eur J Endocrinol. 2000; 143:R11-4.

67. Hataya Y, Akamizu T, Takaya K, et al. A low dose of ghrelin stimulates growth hormone (GH) release synergistically with GH-releasing hormone in humans. J Clin Endocrinol Metab. 2001; 86:4552-5.

68. Wren AM, Small CJ, Abbott CR, et al. Ghrelin causes hyperphagia and obesity in rats. Diabetes. 2001; 50:2540-7.

69. Wren AM, Seal LJ, Cohen MA, et al. Ghrelin enhances appetite and increases food intake in humans. J Clin Endocrinol Metab. 2001; 86:5992-5.

70. Cummings DE, Weigle DS, Frayo RS, et al. Plasma ghrelin levels after diet-induced weight loss or gastric bypass surgery. N Engl J Med. 2002; 346:1623-30.

71. Svensson J, Lonn L, Jansson J-O, et al. Two-Month treatment of obese subjects with the oral growth hormone (GH) secretagogue MK-677 increases GH secretion, fat-free mass, and energy expenditure. J Clin Endocrinol Metab. 1998; 83:362-9.

72. Traebert M, Riediger T, Whitebread S, Scharrer E, Schmid HA. Ghrelin acts on leptin-responsive neurones in the rat arcuate nucleus. J Neuroendocrinol. 2002; 14:580-6.

73. Riediger T, Traebert M, Schmid HA, Scheel C, Lutz TA, Scharrer E. Site-specific effects of ghrelin on the neuronal activity in the hypothalamic arcuate nucleus. Neurosci Lett. 2003; 341:151-5.

74. Cowley MA, Smith RG, Diano S, et al. The distribution and mechanism of action of ghrelin in the CNS demonstrates a novel hypothalamic circuit regulating energy homeostasis. Neuron. 2003; 37:649-61.

75. Smith RG, Sun Y, Paylor R, Paschali A. GH secretagogues and neurodegenerative diseases. Program of the 4th International Symposium on Growth Hormone Secretagogues, Clearwater, FL, USA, 2002 (abstract).

76. Fanselow MS. Conditioned and unconditional components of post-shock freezing. Pavlov J Biol Sci. 1980; 15:177-82.

77. Fanselow MS, Tighe TJ. Contextual conditioning with massed versus distributed unconditional stimuli in the absence of explicit conditional stimuli. J Exp Psychol Anim Behav Process. 1988; 14:187-99.

78. Crawley JN. What's Wrong With My Mouse? New York: Wiley-Liss. 2000.

79. Paylor R, Tracy R, Wehner J, Rudy JW. DBA/2 and C57BL/6 mice differ in contextual fear but not auditory fear conditioning. Behav Neurosci. 1994; 108:810-7.

80. Rigamonti AE, Pincelli AI, Corra B, et al. Plasma ghrelin concentrations in elderly subjects: comparison with anorexic and obese patients. J Endocrinol. 2002; 175:R1-5.

Chapter 10

GHRELIN AND TUMORS

Mauro Papotti, Corrado Ghè[1], Marco Volante & Giampiero Muccioli[1]
Departments of Pathology and [1]Pharmacology, University of Turin, Turin, Italy

Abstract: Ghrelin production was originally demonstrated in human endocrine cells of the gastric body and of the pituitary gland, but was rapidly shown to occur also in several other normal and neoplastic cells. Most endocrine tumors were found to contain ghrelin, as detected by both immunohistochemistry or mRNA analysis. Not only pituitary adenomas and gastro-entero-pancreatic carcinoids produce ghrelin, but also pulmonary carcinoids and thyroid tumors. In addition, we have shown here that nonendocrine lung, breast, colorectal, prostatic, and pancreatic carcinomas may produce ghrelin, as well. The significance of ghrelin expression in human tumors is poorly understood, but it is of interest that concurrent expression of ghrelin binding sites also occurs in many of the above listed neoplasms. Although only one ghrelin receptor has been cloned so far (the growth hormone secretagogue receptor), data from binding studies indicate that a family of ghrelin receptors probably exists in human tissues, including tumors. These receptors share the ability to bind acylated and nonacylated ghrelin, as well as synthetic analogs of the GH secretagogue family. In fact, *in vitro* experiments demonstrated the ability of these compounds to displace radiolabeled ghrelin from binding sites and also their antiproliferative effects on several human tumor cell lines. These findings open interesting perspectives in the control of tumor cell growth using synthetic ghrelin analogs. However, a complete mapping of the ghrelin receptor distribution in human tissues and, above all, tumors is necessary, together with the validation of *in vitro* data on *in vivo* models, to better define the effects of different ghrelin analogs on human tumor growth.

Key words: GH Secretagogue, GHS receptor, human tumors, cell growth

1. INTRODUCTION

Ghrelin production was originally demonstrated in human endocrine cells of the gastric body and of the pituitary gland (1). Subsequently, this hormone was identified in several other endocrine and nonendocrine cell populations

(2-4). The apparent specific expression of ghrelin by selected cell types prompted an analysis of the corresponding endocrine tumors in the pituitary and gastrointestinal tract. The first demonstrations of ghrelin expression by a tumor included studies on pituitary adenomas (5), gastric carcinoids (6) and intestinal or pancreatic endocrine tumors (6,7). The list of ghrelin-expressing tumors is nowadays much longer, thanks to several contributions appeared in the literature. Many findings obtained by different groups are comparable, but there are also discrepancies. A possible explanation is the different method employed to detect ghrelin within tumor cells. Currently employed technical procedures will therefore be briefly reviewed.

2.   METHODS EMPLOYED TO DEMONSTRATE GHRELIN

Ghrelin can be detected at both gene and protein levels by means of *in situ* hybridization (ISH), Northern blotting or reverse transcriptase-polymerase chain reaction (RT-PCR), and by means of immunohistochemistry (IHC).

2.1   Immunohistochemistry

Ghrelin is a 28 amino acid peptide derived from cleavage of a longer precursor peptide of 117 amino acidic residues (pre-pro-ghrelin). The cleaved ghrelin sequence is comprised between residue 24 and residue 51, and includes a serine in position 26 (Ser 3) which is modified by a unique n-octanoylated chain. A rabbit polyclonal antibody to human ghrelin is commercially available (Phoenix Peptide, USA). According to current manufacturer's instruction it was apparently raised against the whole ghrelin sequence. However, it has to be noted that the original information provided in year 2000 in the Phoenix Peptide web site, referred to Kojima's paper in Nature (1) and stated that the antibody (code H-031-30) was raised against the carboxy-terminal fragment 13-28 and was therefore capable of recognizing both the acylated and des-acylated forms of ghrelin. Two other antibodies to ghrelin fragments (12 and 15 amino acid residues, respectively) were raised in chicken (Alpha Diagnostic International, USA) but have not yet been tested in IHC. Some authors developed a polyclonal antiboby to N-terminal or C-terminal ghrelin sequences for IHC and immuno-electron microscopy on human stomach and gastrointestinal endocrine tumors (8) or for immunoassays (5,9; see below). The rabbit antibody to ghrelin works on paraformaldehyde, formalin-fixed and also on alcohol-fixed cells and tissues and does not require any antigen retrieval pre-treatment. The working dilution is to be set in each laboratory using gastric body mucosa as a positive control. In most tumors tested, a dilution of 1/200 up to 1/600 provided best results, in terms of acceptable background and intensity of staining (1,2,5-7).

2.2 Radioimmunoassay procedures

The two major shortcomings of ghrelin IHC are the lack of a protein content quantification and the inability to differentiate between the two forms of the natural hormone, namely octanoylated and des-octanoylated ghrelin. This distinction has biological significance, as the octanoyl residue in the third N-terminal serine appears to be essential for ghrelin biological activities. A radioimmunoassay (RIA) on tumor tissue homogenates has been proposed by several authors (5,9) to define the ghrelin content and type. It is based on the specificity of two different antibodies raised against positions 1-11 (N-terminal) and 13-28 (C-terminal) of ghrelin protein. The former specifically recognizes the octanoylated portion of ghrelin and does not recognize the des-octanoylated protein, while the latter is unable to distinguish between the two. The comparison of high pressure liquid chromatography data using the two different antibodies gives quantitative data on the presence of the protein (usually expressed as fmol/ml or pmol/g of protein extract), and the relative proportion of the biologically active isoform.

2.3 Gene expression analysis

Based on the cloned sequence of the gene (1), several antisense nucleotide sequences were designed to reveal the specific ghrelin mRNA. This was performed with either Northern blotting (1) or ISH (6,7) or RT-PCR. In this latter case, competitive RT-PCR (10) or real time RT-PCR (5) or non-quantitative RT-PCR (1,6,7) were employed. We refer to the original publications for technical details. Since ghrelin mRNA has been widely reported in normal tissues, including blood vessels (11), the demonstration of ghrelin in a tumor cell population needs specific cellular localization of the mRNA/protein by means of ISH or IHC, or of a quantitative PCR reaction. RT-PCR alone may not provide reliable information on the true expression of ghrelin by tumor cells.

3.   GHRELIN PRODUCTION BY ENDOCRINE TUMORS

3.1   Pituitary tumors

Two studies analysed ghrelin expression in pituitary adenomas (5,10). Ghrelin mRNA was detected in all tumor sample tested (92 cases). The hormone was demonstrated with both RT-PCR and IHC. No preferential expression of ghrelin was found in any pituitary adenoma subtype. However, by quantitative measurements of the ghrelin mRNA molecules, it was shown that ghrelin expression was relatively low in corticotroph adenomas and relatively high in somatotroph tumors, as compared with that of normal pituitary tissue (5). Kim and coworkers (10) found the highest ghrelin expression in nonfunctioning adenomas and the lowest in prolactinomas.

3.2   Carcinoid tumors

Gastric body carcinoids were found to produce ghrelin in a relatively high percentage of cases (12/16 and 25/33, or 75%, in two series) (6,8). Most of these cases were associated with chronic atrophic gastritis and were probably of the entero-chromaffin-like (ECL) type, as commonly observed in that location. Although the putative ghrelin producing cell of the stomach is the X/A-like cell (2,8), the corresponding endocrine tumors are exceedingly rare and it is likely that ghrelin is "ectopically" expressed by conventional ECL-carcinoids. The extent of ghrelin expression was either focal or diffuse, as confirmed by several different techniques revealing either the specific mRNA (ISH) or the protein (IHC). It is of interest that peritumoral hyperplastic endocrine cells (in the context of atrophic gastritis) were also intensely expressing ghrelin. Antral carcinoids were generally unreactive. Intestinal carcinoids, conversely, were immunoreactive for ghrelin in only a fraction of cases (27%) (6). These figures are related to immunolocalisation findings. When the more sensitive RT-PCR technique was employed, the vast majority of endocrine tumors turned out to be positive for ghrelin mRNA. Pancreatic well-differentiated endocrine tumors were also found to express ghrelin in variable percentages ranging from 25% (1/4) (12) to 40% (6/15 cases) (8) and 81% (22/27cases) (7), thus confirming the very preliminary data (based on real time RT-PCR) of Korbonits and coworkers (5) who found a high copy number of ghrelin mRNA molecules in 3 pancreatic insulinomas, but not in the single case of gastrinoma. The presence of ghrelin in pancreatic endocrine tumors is not surprising since several groups documented the presence of ghrelin in fetal (8,13) and adult islets of Langerhans. In the latter case, discrepant data appeared in the literature regarding the cell type producing ghrelin. Glucagon-producing alpha cells were proposed as the cell type coexpressing ghrelin in rat pancreas (14). Our group showed a co-expression of insulin and ghrelin in human beta cells (7). Very recently, the existence of a separate cell type, the "ghrelin cell", was proposed in the pancreatic islets (15). Ghrelin-producing cells were much more represented in the fetal pancreas, but persisted occasionally in the adult, either within the islet or scattered in the duct wall (15). Irrespective of the source of ghrelin in normal conditions, endocrine tumors are not apparently showing any preferential ghrelin expression by the different functioning pancreatic tumors. Ghrelin production was observed in insulinomas, glucagonomas, gastrinomas and somatostatinomas (7). Finally, neuroendocrine tumors of the lung have been found to produce ghrelin. Rindi and coworkers (8) detected immunoreactive ghrelin in 4 out of 8 neuroendocrine lung tumors, whereas we were able to document the specific mRNA expression in 8/9 cases, in the absence of immunohistochemically proven protein expression (unpublished results). Ghrelin production was documented also in normal endocrine cells of the lung, particularly during fetal life (13). The significance of this finding in both normal and neoplastic conditions is currently not known. All the above findings were referred to well-differentiated endocrine tumors (i.e. carcinoids). It is however known that a spectrum of endocrine tumors exists, including also highly aggressive, small cell poorly differentiated tumors. These latter were found to lack ghrelin expression in the case of gastric (0/4 cases) (8) or pancreatic (0/1 case) (7) locations and to contain ghrelin mRNA in one of two poorly differentiated primary pulmonary carcinomas (unpublished results).

3.3   Thyroid tumors

In 2001, Kanamoto and coworkers (9) reported that the medullary thyroid carcinoma TT cell line contained immunoreactive ghrelin and its corresponding mRNA. Three cases of human medullary carcinoma were also found to produce ghrelin, as detected by a RIA specifically designed to reveal both acylated and des-acylated ghrelin. Two cases of papillary thyroid carcinoma tested in parallel did not show a high content of ghrelin, which was in the range of that of normal thyroid (9). Recently, we have shown that ghrelin is widely expressed by follicular-derived thyroid tumors, as detected by IHC and/or RT-PCR (16). Follicular adenomas focally expressed ghrelin. Malignant follicular tumors diffusely expressed ghrelin at both protein and mRNA levels. The extent of ghrelin expression was not related to tumor grade, the percentage of tumors which tested positive being 75% of papillary, 90% of follicular, 100% of poorly differentiated and 50% of anaplastic carcinomas. The significance of ghrelin production by thyroid tumor cells is poorly understood, but the reported presence of ghrelin/GH secretagogue binding sites in thyroid tumors (17) suggests that the hormone may act directly on the tumor cells via specific receptors. An inhibitory effect on tumor cell growth of thyroid carcinoma cell lines has in fact been demonstrated (16,18) (see also below).

4.   GHRELIN PRODUCTION BY NON-ENDOCRINE TUMORS

Information regarding ghrelin production by non-endocrine tumors is largely confined to the analysis of human tumor cell lines (see below). Ghrelin mRNA and/or protein was demonstrated in four prostate carcinoma cell lines (19) and in six human leukemic cell lines (20). Preliminary data from our laboratory indicate that ghrelin mRNA is detectable in several non-endocrine human tumors by means of RT-PCR, including 5/10 pulmonary squamous cell or adenocarcinomas, 4/4 pancreatic ductal adenocarcinomas, 4/4 gastric and 4/4 colorectal adenocarcinomas, 8/10 prostatic adenocarcinomas, 1/6 hepatocellular carcinomas, and 8/8 breast carcinomas (Figure 10-1).

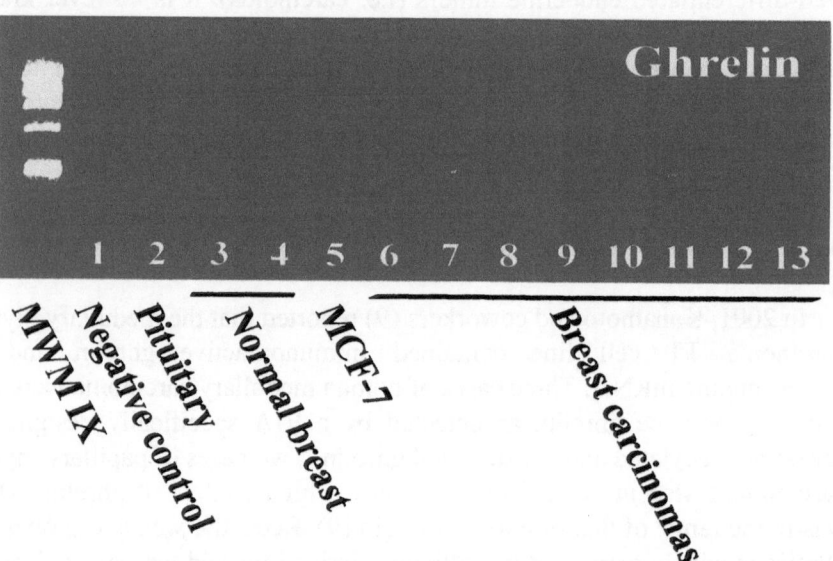

Figure 10-1. Ghrelin mRNA expression as detected by means of RT-PCR in normal breast tissue. MCF-7 breast carcinoma cell line and breast carcinoma tissue samples.

The corresponding protein, as revealed by IHC on formalin-fixed archival material, was however only occasionally demonstrated, indicating that the amount of stored peptide in the tumor cells is probably not very high.

5. SIGNIFICANCE OF GHRELIN EXPRESSION

Current data indicate that ghrelin is widely expressed by human tumors, both of the endocrine and, apparently, also of the non-endocrine type (Table 10-1). These finding parallel recent observations regarding the widespread occurrence of ghrelin in normal human tissues (11) and in fetal tissues (8,13). The functional role of ghrelin in such tumors and tissues is not completely understood and it is strongly related to the presence of functional ghrelin binding sites. The only currently known ghrelin receptors are Growth Hormone Secretagogue Receptor (GHS-R) 1a and 1b. The functional implications of increased ghrelin expression will therefore be dealt with after reviewing the distribution of ghrelin binding sites in human tumors and cell lines.

6. GHRELIN BINDING SITES IN TUMORS

Ghrelin binding sites have been known for several years, even before the identification of this hormone, since ghrelin receptors are now known to be shared by synthetic peptides called Growth Hormone Secretagogues (GHS) (21). These compounds exerted their growth hormone (GH)-releasing activity by binding to membrane receptors apparently devoid of a natural ligand (orphan receptors). The discovery of ghrelin filled the gap as both ghrelin and synthetic analogues compete for these same binding sites. Only one receptor, the GHS-R, has been cloned, in its two splice variants labeled GHS-R 1a and 1b. Binding studies based on displacement capacity of radiolabeled synthetic peptidyl GHS ([^{125}I]Tyr-Ala-hexarelin) suggested that some normal and tumor tissues bind ghrelin and synthetic GHS, but lack the specific GHS-R 1a mRNA. This strongly suggests that other receptor subtypes may be responsible for the observed binding. Therefore the identification of ghrelin binding sites needs a combination of different techniques, in order to assess the presence, the type and the cellular localisation of the receptor.

Table 10-1. Ghrelin expression in human normal and tumoral tissue.

Organ	Source	Ghrelin	Method	Ref.
Lung	Normal fetal tissue	+ (20/20)	ICC	13
	Normal infant/adult tissue	+ (15/20)	ICC/RT-PCR	13
	NE lung tumors	+ (9/11)	RT-PCR	α
		+ (4/8)	ICC	8
	Non-NE lung tumors	+ (5/10)	RT-PCR	α

Pancreas	Normal islet cells	+ (7/7)	ICC/RT-PCR	7
	Endocrine tumors	+ (22/28)	ICC/RT-PCR	7
		+ (9/11)	ICC	8
		+ (9/11)	RT-PCR	5
	Exocrine tumors	+ (4/4)	RT-PCR	β
Gastro-Intestinal tract	Endocrine gastric tumors	+ (12/16)	ICC/RT-PCR	6
		+ (25/33)	ICC	8
	Endocrine intestinal tumors	+ (11/18)	ICC/RT-PCR	6
	Non-endocrine gastric tumors	+ (4/4)	RT-PCR	β
	Non-endocrine intestinal tumors	+ (4/4)	RT-PCR	β
Prostate	Adenomatous hyperplasia	+ (10/11)	RT-PCR	α
	Adenocarcinoma	+ (8/10)	RT-PCR	α
	Prostate ca. cell lines (ALVA41, DU145, LNCaP, PC3)	+ (4/4)	ICC/RT-PCR	19
Liver	Normal liver	- (0/2)	RT-PCR	α
	Hepatocellular ca.	+ (1/6)	RT-PCR	α
Thyroid	Normal thyroid tissue	+ (3/10)[a]	ICC/RT-PCR	16
	Follicular adenoma	+ (13/18)	ICC/RT-PCR	16
	Papillary ca.	+ (9/12)	ICC/RT-PCR	16
		- (0/2)	RIA	9
	Follicular ca.	+ (9/10)	ICC/RT-PCR	16
	Poorly differentiated ca.	+ (8/8)	ICC/RT-PCR	16
	Anaplastic ca.	+ (3/6)	ICC/RT-PCR	16
	Medullary ca.	+ (3/3)	RIA	9
	Papillary ca. (N-PAP) cell line	+	ICC	16
	Anaplastic ca. (ARO) cell line	+	ICC	16
	Medullary ca. (TT) cell line	+	ICC/RT-PCR	9
Breast	Normal parenchima	- (0/2)	RT-PCR	α
	Carcinomas	+ (8/8)	RT-PCR	α
	Breast ca. cell lines (MCF7)	-	RT-PCR	α
Lymphoid tissue	Non-Hodgkin lymphoma	+ (3/3)	RT-PCR	α
	Hodgkin lymphoma	+ (3/4)	RT-PCR	α
	T leukemic (Jurkat,Hut78) cell line	+ (2/2)	RT-PCR	20
	B leukemic (Raji,Daudi) cell line	+ (2/2)	RT-PCR	20
	Myeloid leukemic (K-562,HL-60) cell line	+ (2/2)	RT-PCR	20

Pituitary	GH secreting adenoma	+ (13/13)	coRT-PCR	10
		+ (22/22)	ReT RT-PCR	5
	PRL secreting adenoma	+ (4/4)	coRT-PCR	10
		+ (4/4)	ReT RT-PCR	5
	Gonadotropin secreting	+ (5/5)	coRT-PCR	10
	adenoma	+ (5/5)	ReT RT-PCR	5
		+ (2/2)	coRT-PCR	10
	ACTH secreting adenoma	+ (12/12)	ReT RT-PCR	5
		+ (2/2)	coRT-PCR	10
	TSH secreting adenoma	+ (4/4) ·	coRT-PCR	10
	Non-functioning adenoma	+ (12/12)	ReT RT-PCR	5

Legend. NE: neuroendocrine; GH: growth hormone; PRL: prolactin; ACTH: adrenocorticotropic hormone; TSH: thyroid stimulating hormone; ca.: carcinoma; +: positive; -: negative; ICC: immunocytochemistry; RT-PCR: reverse transcriptase-polymerase chain reaction; RIA: Radio Immuno Assay; coRT-PCR: competitive RT-PCR; ReT RT-PCR: Real Time RT-PCR; [a]: only positive for RT-PCR; α: Papotti (unpublished results); β: Papotti and Muccioli (unpublished results).

7. METHODS TO DEMONSTRATE GHRELIN BINDING SITES

Originally, the presence of binding sites to natural (ghrelin) and synthetic (e.g. hexarelin) GHS was demonstrated by *in vitro* binding assays, using tissue homogenates and radiolabeled compounds. This procedure is highly sensitive and specific, and allows definition of the receptor affinity for the given ligands, but does not allow identification of the receptor molecule. Only one ghrelin receptor has been isolated so far, the GHS-R, which has two isoforms referred to as GHS-R 1a and 1b (21). However, as mentioned above, recent data showed that this is not the only existing ghrelin receptor, since in several tissues, mostly peripheral, the presence of specific binding sites for ghrelin and synthetic GHS was not paralleled by the presence of the specific GHS-R 1a mRNA. The different technical procedures are therefore identifying partially different receptors.

7.1 Binding studies

Several methods have been described in the literature. *In vitro* receptor autoradiography studies using [125I]His[9]-ghrelin as a ligand revealed the presence of specific binding sites in cardiovascular tissues (22). Using [125I]Tyr[4]-ghrelin as radiotracer on tissue membranes (and saturation and competitive displacement experiments), our group demonstrated the presence of a single class of binding sites for ghrelin in the human hypothalamus and pituitary gland (23). These high affinity (K_d 0.29-0.56 nM) receptors are capable of binding with high affinity also synthetic peptidyl (hexarelin) and nonpeptidyl (MK-0677) GHS (23). Before the discovery of ghrelin, radiolabeled synthetic nonpeptidyl ([^{35}S-MK]-0677) and peptidyl GHS ([^{125}I]Tyr-Ala-hexarelin) were used to reveal specific binding sites in the brain, pituitary gland and in a wide range of endocrine and non-endocrine peripheral human tissues and tumors (see below) (17,24-28). GHS-R different from the known GHS-R 1a were also demonstrated in the pituitary and in the heart by means of a specific radioreceptor assay (25,29) and a photo-affinity labeling technique using [^{125}I]Tyr-Ala-hexarelin and [^{125}I]benzoyl-L-Phe-Ala-hexarelin, respectively (30,31).

7.2   GHS-R gene expression analysis

Based on the cloned sequence of the GHS-R 1a and 1b genes (21), antisense nucleotide sequences and primers were designed to reveal the specific ghrelin mRNA. This was performed with either Northern blotting or RT-PCR. We refer to the original publications for technical details. Since GHS-R mRNA has been reported in several normal tissues, its demonstration in a tumor cell population needs specific localization of the mRNA/protein in the cells by means of ISH or IHC. RT-PCR alone may not provide reliable information on the true expression of GHS-R by the tumor cells.

7.3   Immunohistochemistry and *in situ* hybridisation

In situ techniques to localize the receptor at the cellular level include ISH and IHC. The former was employed by several authors to detect GHS-R specific mRNA expression in lemur and rat central nervous system (32-36), in the human brain (33) as well as in human pituitary adenomas (37). IHC is a valuable, cheap and specific tool to visualize the receptor protein, having the only drawback of a relatively lower sensitivity, as compared to gene expression analysis. Only one paper appeared in the literature (38) describing a polyclonal antibody to rat GHS-R 1a which was shown to react with tissue extracts from rat hypothalamus, pituitary and stomach in Western blot assays. By using this polyclonal antibody in IHC experiments, the reaction was found to be mainly confined to the cell membrane with a minor staining of the cytoplasm. Two polyclonal antisera are commercially available for IHC and future studies are necessary to map the exact cellular distribution of the GHS-R in human tissues and tumors.

8.   GHRELIN BINDING SITES IN ENDOCRINE TUMORS

8.1   Pituitary tumors

Seven studies appeared dealing with the analysis of ghrelin binding sites in the pituitary gland and pituitary tumors (5,10,37,39-42). Standard or quantitative RT-PCR procedures were employed. Over 200 adenomas were studied. The vast majority (197/220) were found to contain GHS-R mRNA transcripts, regardless of the tumor cell type and hormone produced. Non-functioning adenomas were also shown to express GHS-R mRNA by most authors (5,10,37). The only exception in this regard were the data by Adams and coworkers (39) and the preliminary study of Korbonits and coworkers (41) where no demonstration of GSH-R mRNA could be obtained by RT-PCR in any of their 18 nonfunctioning adenomas. In the two studies (5,10) in which GHS-R and ghrelin expression were compared in the same tumors, a correlation was found between the two, indicating that ligand/receptor interactions may take place in the same tumor and possibly in the same cells through autocrine or paracrine circuits.

8.2   Neuroendocrine tumors (carcinoids)

In 1998, Korbonits and coworkers (41) reported that a pancreatic gastrinoma showed levels of GHS-R mRNA expression (as assessed by RT-PCR) similar to those of the normal pituitary. The same authors also found low level expression of GHS-R mRNA in three insulinomas and no detectable expression in a nonfunctioning thymic carcinoid (5). Our group analyzed a relatively large series of 42 gastroentero-pancreatic neuroendocrine tumors (6,7). By means of RT-PCR, 14/28 pancreatic endocrine tumors were shown to contain GHS-R 1b mRNA and half of them also contained type 1a transcripts. All GHS-R 1a positive tumors were insulinomas, while those positive for type 1b included either insulinomas or other functioning and nonfunctioning neoplasms. Of the 14 gastro-intestinal (GI) carcinoids examined in our study, 7 were shown to contain GHS-R 1a and/or 1b (all 7 positive carcinoids being localized in the intestine, rather than in the stomach). All GHS-R-positive carcinoids also expressed ghrelin mRNA, as detected by RT-PCR. Similarly, neuroendocrine tumors of the lung were found to contain GHS binding sites (28,42). Specific binding values of radiolabeled Tyr-Ala-hexarelin to tumor membrane homogenates was detected in 8 endocrine tumors, the values of binding being similar to those of non-tumoral lung. Much higher binding values were observed in non-neuroendocrine lung carcinomas (see below). When the same cases were analyzed for GHS-R mRNA expression, using RT-PCR, only half of the tested cases contained the appropriate mRNA, confirming that in peripheral tissues additional receptors probably exist, other than GHS-R 1a and 1b (28).

8.3   Thyroid tumors

Specific binding values of radiolabeled Tyr-Ala-hexarelin to tumor membrane homogenates was detected in 30 benign or malignant human thyroid tumors of follicular origin (follicular adenoma and papillary, follicular or anaplastic carcinomas), the values of binding found in benign tumors being similar to those of non-tumoral thyroid parenchyma and those in the case of well differentiated carcinomas much higher (17). No binding was observed in 6 C-cell derived medullary carcinomas. When the specific GHS-R 1a and 1b mRNAs were analyzed, none of 33 follicular-derived tumors was shown to express either form of the receptor, indicating that in the thyroid gland the observed GHS binding must be mediated by a different receptor (16). In normal human thyroid, only 1 of 5 samples was found to contain GHS-R 1b mRNA by standard RT-PCR (16), but Gnanapavan and coworkers (11) found strong expression of both GHS-R 1a and 1b mRNAs using real-time PCR.

9.   GHRELIN BINDING SITES IN NON-ENDOCRINE TUMORS

Specific binding of radiolabeled Tyr-Ala-hexarelin to tumor membrane homogenates was detected in 24 of 24 breast carcinomas (ductal and lobular types), but not in 10 benign breast disease (27), and in 12 of 12 non-small cell lung carcinomas (28). In addition, preliminary data from our laboratory indicate that also pancreatic ductal adenocarcinomas (3 of 3), hepatocellular carcinomas (4 of 4), gastric and colorectal adenocarcinomas (8 of 8) contain specific GHS binding sites. RT-PCR for the specific GHS-R 1a and 1b mRNAs performed in the same samples, however, could not demonstrate the receptor in none of the hepatocellular (0 of 6), pulmonary (0 of 12) or breast (0 of 8) carcinomas studied (28; Papotti, unpublished results). By using RT-PCR, GHS-R 1a and 1b mRNA was not found in 7 Hodgkin or non-Hodgkin lymphomas and type 1b only was detected in 1 out of 10 prostatic adenocarcinomas, as opposed to 5 of 13 benign prostatic hyperplasias (unpublished results). All data obtained by other authors on the GHS-R expression in non-endocrine tumors derive from *in vitro* models (see below). GHS-R mRNA was detected by RT-PCR in 4 prostatic carcinoma cell lines (19), in the HepG2 hepatoma cell line (43) and in 6 leukemic cell lines (both

T or B lymphoid and myeloid cell lines) (20). In contrast, this same receptor type was not found in the lung carcinoma CALU-1 cell line, nor in 3 hormone dependent or independent breast cancer cell lines (27,28).

10. SIGNIFICANCE OF GHRELIN RECEPTOR EXPRESSION

Current data indicate that ghrelin binding sites are widely expressed in human tissues and in several endocrine and non-endocrine tumors (Table 10-2). As discussed above, this finding is paralleled by the widespread occurrence of ghrelin in the same tissues. The exact functional significance of GHS/ghrelin receptors in human tissues is not completely understood. Recent data indicate that, apart from mediating the hormonal activities promoted by ghrelin, other non-endocrine functions may be associated with the presence of GHS/ghrelin receptors, including regulatory effects on tumor cell growth.

Table 10-2. Ghrelin binding sites in human normal and tumoral tissues.

Organ	Source	GHS-R	Method	Ref.
Lung	Normal infant/adult tissue	+ (2/20)	RT-PCR	13
		- (0/3)	RT-PCR	28
		+ (10/10)	Binding	28
	NE lung tumors	+ (17/21)	RT-PCR	42
		+ (8/11)	RT-PCR	α
		+ (3/6)	RT-PCR	28
		+ (8/8)	Binding	28
	Non-NE lung tumors	+ (1/9)	RT-PCR	42
		- (0/12)	RT-PCR	28
		+ (12/12)	Binding	28
	Lung ca. (CALU-1) cell line	-	RT-PCR	28
		+	Binding	28
Pancreas	Normal islet cells	- (0/5)	RT-PCR	7
	Endocrine tumors	+ (14/28)	RT-PCR	7
		+ (1/4)	RT-PCR	5, 41
	Non-endocrine tumors	+ (3/3)	Binding	β
Gastro-Intestinal tract	Endocrine gastric tumors	- (0/3)	RT-PCR	6
	Non-endocrine gastric tumors	+ (4/4)	Binding	β
	Endocrine intestinal tumors	+ (7/11)	RT-PCR	6
	Non-endocrine intestinal tumors	+ (4/4)	Binding	β

Prostate	Adenomatous hyperplasia	+ (5/13)	RT-PCR	α
	Adenoca.	+ (1/10)	RT-PCR	α
	Prostate ca. cell lines			
	(ALVA41,DU145,LNCaP,PC3)	+ (4/4)	ICC/RT-PCR	19
Liver	Normal liver	+ (1/1)	RT-PCR	43
		- (0/2)	RT-PCR	α
	Hepatocellular ca.	- (0/6)	RT-PCR	α
		+ (4/4)	Binding	β
	Hepatoma (HepG2) cell line	+	RT-PCR	43
Thyroid	Normal thyroid tissue	- (0/5)	Binding	17
		+ (1/5)	RT-PCR	16
	Follicular adenoma	+ (9/9)	Binding	17
		- (0/12)	RT-PCR	16
	Papillary ca.	+ (13/13)	Binding	17
		- (0/10)	RT-PCR	16
	Follicular ca.	+ (4/4)	Binding	17
		- (0/6)	RT-PCR	16
	Poorly differentiated ca.	+ (2/2)	Binding	17
		- (0/1)	RT-PCR	16
	Anaplastic ca.	+ (2/2)	Binding	17
		- (0/4)	RT-PCR	16
	Medullary ca.	- (0/6)	Binding	17
	Papillary ca. (N-PAP) cell line	-	RT-PCR	16
	Anaplastic ca. (ARO) cell line	-	RT-PCR	16
Breast	Normal parenchima	- (0/6)	Binding	27
	Benign tumors	- (0/4)	Binding	27
	Carcinomas	+ (24/24)	Binding	27
		- (0/8)	RT-PCR	α
	Breast ca. cell lines	+ (3/3)	Binding	27
	(MDA,MCF7,T47D)	- (0/3)	RT-PCR	27
Lymphoid tissue	Non-Hodgkin lymphoma	- (0/3)	RT-PCR	α
	Hodgkin lymphoma	- (0/4)	RT-PCR	α
	T leukemic (Jurkat,Hut78) cell line	+ (2/2)	RT-PCR	20
	B leukemic (Raji,Daudi) cell line	+ (2/2)	RT-PCR	20
	Myeloid leukemic (K-562,HL-60) cell line	+ (2/2)	RT-PCR	20

Pituitary	GH secreting adenoma	+ (5/5)	RT-PCR	42
		+ (8/8)	RT-PCR	41
		+ (6/6)	RT-PCR	39
		+ (10/10)	RT-PCR	40
		+ (6/6)	RT-PCR	37
		+ (13/13)	coRT-PCR	10
		+ (22/22)	ReT RT-PCR	5
	PRL secreting adenoma	+ (2/3)	RT-PCR	42
		+ (2/4)	RT-PCR	41
		+ (3/3)	RT-PCR	39
		+ (6/6)	RT-PCR	37
		+ (4/4)	coRT-PCR	10
		+ (4/4)	ReT RT-PCR	5
	Gonadotropin secreting adenoma	- (0/1)	RT-PCR	41
		+ (3/9)	RT-PCR	40
		+ (2/4)	RT-PCR	37
		+ (5/5)	coRT-PCR	10
		+ (5/5)	ReT RT-PCR	5
	ACTH secreting adenoma	+ (18/18)	RT-PCR	42
		+ (13/18)	RT-PCR	41
		+ (3/4)	RT-PCR	40
		+ (2/3)	RT-PCR	37
		+ (2/2)	coRT-PCR	10
		+ (12/12)	ReT RT-PCR	5
	TSH secreting adenoma	+ (1/1)	RT-PCR	40
		- (0/1)	RT-PCR	37
		+ (2/2)	coRT-PCR	10
	Non-functioning adenoma	+ (1/2)	RT-PCR	42
		- (0/10)	RT-PCR	41
		- (0/8)	RT-PCR	39
		+ (1/5)	RT-PCR	37
		+ (4/4)	coRT-PCR	10
		+ (12/12)	ReT RT-PCR	5

Legend. NE: neuroendocrine; GH: growth hormone; PRL: prolactin; ACTH: adrenocorticotropic hormone; TSH: thyroid stimulating hormone; ca.: carcinoma; +: positive; -: negative; ICC: immunocytochemistry; RT-PCR: reverse transcriptase-polymerase chain reaction; RIA: Radio Immuno Assay; coRT-PCR: competitive RT-PCR; ReT RT-PCR: Real Time RT-PCR; α: Papotti (unpublished results); β: Papotti and Muccioli (unpublished results).

11.   FUNCTIONAL ROLE OF GHRELIN/GHS-R IN TUMORS

11.1   Ghrelin in tumor cell line

In 2001, Kanamoto and coworkers (9) described immunoreactive ghrelin and its corresponding mRNA in the medullary thyroid carcinoma TT cell line. Similar results were also obtained in 6 leukemic cell lines, such as T (Jurkat, Hut-78) or B (Raji, Daudi) lymphoid and myeloid (K-562, HL-60) cells (20). Ghrelin mRNA and/or protein was also demonstrated in two androgen-dependent (ALVA41, LNCaP) and two androgen-independent (DU145, PC3) prostatic carcinoma cell lines (19). Preliminary data from our laboratory indicate that ghrelin mRNA and protein are detectable in several non-endocrine human tumor cell lines, including human pancreatic (CAPAN-I and CAPAN-2), gastric (Kato III) and colon (HT-29) adenocarcinoma cell lines. Taken together, these data indicate that ghrelin is widely expressed by different cancer cell lines, both of the endocrine and non-endocrine types. These findings parallel recent observations regarding the widespread occurrence of ghrelin in tumor tissues (see above). The significance of ghrelin expression in neoplastic cell lines is not completely understood. The biological activities of the hormone may be sustained by both acylated and non-acylated forms that cannot be distinguished by the current IHC or ISH procedures. Although the des-octanoylated molecule does not have GH-releasing effects *in vivo*, it represents the largest part of circulating ghrelin and, in cultured H9c2 cardiomyocytes and endothelial cells, has been shown to prevent cell death induced by different pro-apoptotic compounds (44). Therefore, this endogenous ghrelin-derived molecule could have non-endocrine activities including a possible role in modulating the neoplastic cell growth (see below).

11.2   GHS-R in tumor cell lines

GHS-R 1a and 1b have been detected, by RT-PCR or IHC, in ALVA41, LNCaP, DU145 and PC3 human prostatic carcinoma cell lines (19) and in TT medullary thyroid carcinoma cells (18). GHS-R 1a mRNA was also found in rat pituitary adenoma GH_3 cells (39), in the human (HepG2) and rat (H4-II-E) hepatoma cell lines (43), in Jurkat, Hut-78, Raji, Daudi K-562and HL-60 leukemic cell lines (20), and more recently in the pancreatic adenocarcinoma cell line CAPAN-1 (unpublished results). Conversely, GHS-R 1a was not found in the estrogen-dependent (MCF7, T47D) or independent (MDA-MB231) breast cancer cell lines (27), nor in the non-endocrine lung carcinoma cell line CALU-1 (28). However, our recent data indicated that breast and lung cancer cell lines possess specific GHS binding sites distinct from the GHS-R 1a. In breast cancer cells, the specific binding of [^{125}I]Tyr-Ala-hexarelin (Figure 10-2) is inhibited, not only by synthetic (hexarelin, MK0677) or natural (ghrelin) GHS, but even by molecules that are devoid of GH-releasing activity *in vivo*, such as the endogenous des-acylated form of ghrelin and the synthetic peptide EP-80317 (27). By contrast, in the lung CALU-1 carcinoma cells the binding of [^{125}I]Tyr-Ala-hexarelin is inhibited by synthetic peptidyl GHS (hexarelin, GHRP6), but not by ghrelin or the synthetic non-peptidyl GHS MK-0677 (28). The presence of Tyr-Ala-hexarelin binding sites was also demonstrated in some human thyroid carcinoma cell lines of follicular origin, such as the papillary (NPA), follicular (WRO) and anaplastic (ARO) cells (17). Taken together, these data support the hypothesis that different GHS binding site subtypes, distinct from GHS-R 1a and 1b, may exist in tumoral cells, possibly depending on their endocrine or non-endocrine nature, but also on their cell differentiation and embryological origin.

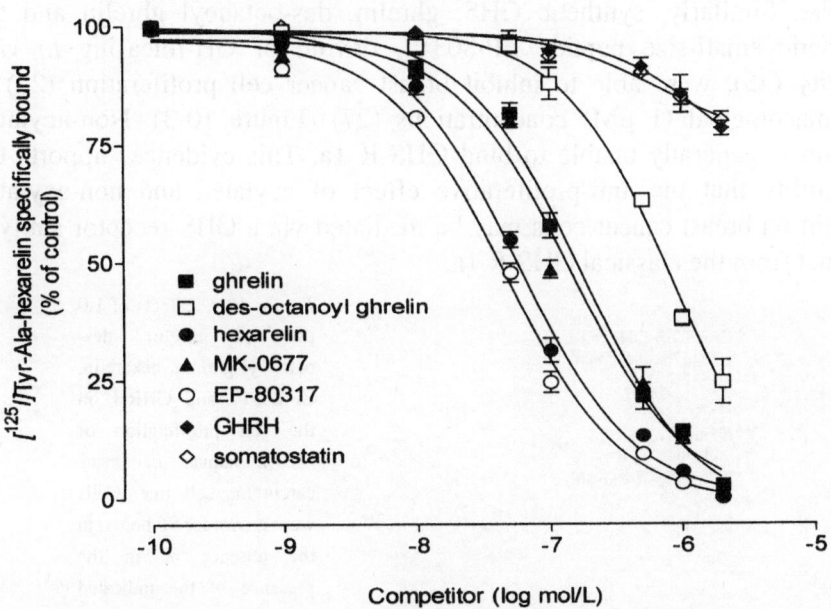

Figure 10- 2. Displacement of [^{125}I]Tyr-Ala-hexarelin from membranes of MCF7 mammary carcinoma cells by different unlabeled competitors. Values are means±standard error of the mean (SEM) of four separate experiments.

11.3   Effects of ghrelin and ghrelin analogs on tumor cell growth

The finding that ghrelin and ghrelin/GHS receptors are co-expressed in several tumors and related cell lines suggests that ghrelin is likely to play an important autocrine/paracrine role in neoplastic conditions. Recent data indicate that, apart from mediating the hormonal activities promoted by ghrelin, other non-endocrine functions may be associated with the presence of ghrelin/GHS receptors, including an inhibitory effect of GHS in thyroid carcinoma cell lines that has been reported by our group (17). In those studies (performed before the discovery of the endogenous GHS-R ligand ghrelin), we observed that synthetic peptidyl and non-peptidyl GHS inhibited [^{3}H]thymidine incorporation stimulated by fetal calf serum and caused growth inhibition of NPA, ARO and WRO thyroid carcinoma cell lines. This effect was evident at the earliest time of treatment (24 hours) and was maximal after 48 hours. More recently, the anti-proliferative effect of synthetic GHS was confirmed in some breast cancer cell lines (MCF7, T47D and MDA-MB231) in which GHS binding sites could be detected by means of [^{125}I]Tyr-Ala-hexarelin labeling, in the absence of the specific GHS-R mRNA. Similarly, synthetic GHS, ghrelin, des-octanoyl ghrelin and the synthetic small-size peptide EP-80317, devoid of GH-releasing *in vivo* activity (45), were able to inhibit breast cancer cell proliferation (27) at pharmacological (1 μM) concentrations (27) (Figure 10-3). Non-acylated ghrelin is generally unable to bind GHS-R 1a. This evidence supports the possibility that the anti-proliferative effect of acylated and non-acylated ghrelin on breast cancer cells may be mediated via a GHS receptor subtype distinct from the classical GHS-R 1a.

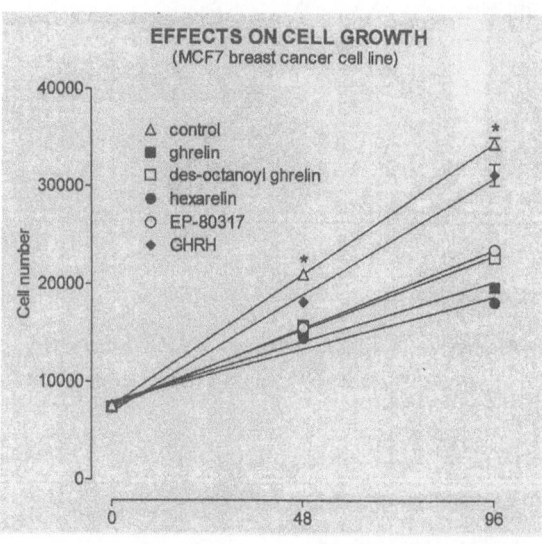

Figure 10-3. Effects of 1.0 μM of ghrelin, des-octanoyl ghrelin, hexarelin, EP-80317 and GHRH on the cell proliferation of MCF7 human mammary carcinoma cell line. Cells were grown for 96 hours, in the absence or in the presence of the indicated compounds and counted every 48 hours. Values are means±SEM of three separate experiments. *p<0.01 (ghrelin, des-octanoyl ghrelin, hexarelin, EP-80317) vs control.

At variance with data in breast cancer cells, we have also shown that DNA synthesis and proliferation of CALU-1 lung carcinoma cells (known to lack the GHS-R1a) is inhibited by peptidyl GHS, but not by non-peptidyl GHS and ghrelin, suggesting that different ghrelin receptors or GHS-R variants exist in non-endocrine tumors (28). However, in human hepatoma cells (HepG2) (known to express GHS-R 1a), ghrelin at concentration ranging from 10 to 100 nM has been reported to stimulate cell proliferation through activation of MAP kinase (43). A similar mitogenic effect has been observed in the PC3 prostatic carcinoma cell line (found to express ghrelin, GHS-R 1a and 1b) after a three-day exposure to physiological ghrelin concentrations (5-10 nM). The dose-related increase in cell number observed in this cell line peaked at 5 nM, decreased at 10 nM and was not observed at higher concentrations (20 nM). Such results on PC3 cells have been confirmed in our laboratories. Using a wide range of ghrelin concentrations (from 1nM to 1μM), we observed that ghrelin exerts a biphasic effect on PC3 cell growth, with a stimulatory activity between 1 and 10 nM and an inhibitory dose-related effect on cell proliferation at doses ranging from 50 nM to 1 μM (unpublished results).

The modulating effects of ghrelin and ghrelin analogs on tumor cell growth further confirm their multiple biological activities (46,47) possibly related to different receptor types or isoforms. The possibility of an anti-neoplastic role of these compounds needs extensive further investigations: a complete map of tumors containing specific and functional receptors is still missing, and, more important, synthetic peptide(s) virtually devoid of remarkable GH-releasing effects are still to be selected and tested in *in vitro* and *in vivo* models.

ACKNOWLEDGEMENTS

The authors wish to thank Prof. F. De Matteis for critically reviewing the manuscript and Drs. E. Allìa, E. Arvat, F. Broglio, P. Cassoni, F. Catapano, R. Deghenghi, T. Marrocco and E. Tarabra for their collaboration in this study project. The present review was supported by grants from the Italian Ministry of Education and University (MIUR, Rome) (ex-60% to GM and MP and Cofin 2002 to GM), the Associazione Italiana per la Ricerca sul Cancro (AIRC, Milan), and the Fondazione per lo Studio delle Malattie Endocrine e Metaboliche (SMEM, Turin).

REFERENCES

1. Kojima M, Hosoda H, Date Y, Nakazato M, Matsuo H, Kangawa K. Ghrelin is a growth hormone-releasing acylated peptide from stomach. Nature. 1999; 402:656-60.
2. Date Y, Kojima M, Hosoda H, et al. Ghrelin, a novel growth-hormone-releasing acylated peptide, is synthesized in a distinct endocrine cell type in the gastro-intestinal tracts of rats and humans. Endocrinology. 2000; 141:4255-61.
3. Mori K, Yoshimoto A, Takaya K, et al. Kidney produces a novel acylated peptide, ghrelin. FEBS Lett. 2000; 486:213-6.
4. Gualillo O, Caminos J, Blanco M, et al. Ghrelin, a novel placental-derived hormone. Endocrinology. 2001; 142:788-94.
5. Korbonits M, Bustin SA, Kojima M, et al. The expression of the growth hormone secretagogue receptor ligand ghrelin in normal and abnormal human pituitary and other neuroendocrine tumors. J Clin Endocrinol Metab. 2001; 86:881-7.
6. Papotti M, Cassoni P, Volante M, Deghenghi R, Muccioli G, Ghigo E. Ghrelin-producing edocrine tumors of the stomach and intestine. J Clin Endocrinol Metabol. 2001; 86:5052-9.
7. Volante M, Allia E, Gugliotta P, et al. Expression of ghrelin and of the GH secretagogue receptor by pancreatic islet cells and related endocrine tumors. J Clin Endocrinol Metab 2002; 87:1300-8.
8. Rindi G, Savio A, Torsello A, et al. Ghrelin expression in gut endocrine growths. Histochem Cell Biol. 2002; 117:521-5.
9. Kanamoto N, Akamizu T, Hosoda H, et al. Substantial production of ghrelin by a human medullary thyroid carcinoma cell line. J Clin Endocrinol Metabol. 2001; 86:4984-90.
10. Kim K, Arai K, Sanno N, Osamura RY, Teramoto A, Shibasaki T. Ghrelin and growth hormone (GH) secretagogue receptor (GHS-R) mRNA expression in human pituitary adenomas. Clin Endocrinol. 2001; 54:759-68.
11. Gnanapavan S, Kola B, Bustin SA, et al. The tissue distribution of the mRNA of ghrelin and subtypes of its receptor, GHS-R, in humans. J Clin Endocrinol Metab. 2002; 87:2988-91.
12. Iwakura H, Hosoda K, Doi R, et al. Ghrelin expression in islet cell tumors: augmented expression of ghrelin in a case of glucagonoma with multiple endocrine neoplasm type I. J Clin Endocrinol Metab. 2002; 87:4885-8.
13. Volante M, Fulcheri E, Allia E, Cerrato M, Pucci A, Papotti M. Ghrelin expression in fetal, infant, and adult human lung. J Histochem Cytochem. 2002; 50:1013-21.
14. Date Y, Nakazato M, Hashiguchi S, et al. Ghrelin is present in pancreatic alpha-cells of humans and rats and stimulates insulin secretion. Diabetes. 2002; 51:124-9.
15. Wierup N, Svensson H, Mulder H, Sundler F. The ghrelin cell: a novel developmentally regulated islet cell in the human pancreas. Regul Pept. 2002; 107:63-9.
16. Volante M, Allia E, Fulcheri E, et al. Ghrelin in fetal thyroid and follicular tumors and cell lines: expression and effects on tumor growth. Am J Pathol. 2003; 645-54.
17. Cassoni P, Papotti M, Catapano F, et al. Specific binding sites for synthetic growth hormone secretagogues in nontumoral and neoplastic human thyroid tissue. J Endocrinol. 2000; 165:139-46.
18. Cassoni P, Muccioli G, Marrocco T, et al. Cortistatin-14 inhibits cell proliferation of human thyroid carcinoma cell lines of both follicular and parafollicular origin. J Endocrinol Invest. 2002; 25:362-8.

19. Jeffery PL, Herington AC, Chopin LK. Expression and action of the ghrowth hormone releasing peptide ghrelin and its receptor in prostate cancer cell lines. J Endocrinol. 2002; 172:R7-11.

20. Hattori N, Saito T, Yagyu T, Jiang BH, Kitagawa K, Inagaki C. GH, GH receptor, GH secretagogue receptor, and ghrelin expression in human T cells, B cells, and neutrophils. J Clin Endocrinol Metab. 2001; 86:4284-91.

21. Howard AD, Feighner SD, Cully DF, et al. A receptor in pituitary and hypothalamus that functions in growth hormone release. Science. 1996; 273:974-7.

22. Katugampola SD, Pallikaros Z, Davenport AP. [^{125}I-His^9]-ghrelin, a novel radioligand for localizing orphan GHS receptors in human and rat tissue: up-regulation of receptors with atherosclerosis. Br J Pharmacol. 2001; 134:143-9.

23. Muccioli G, Papotti M, Locatelli V, Ghigo E, Deghenghi R. Binding of 125I-labeled ghrelin to membranes from human hypothalamus and pituitary gland. J Endocrinol Invest. 2001; 24:RC7-9.

24. Pong SS, Chaung LY, Dean DC, Nargund RP, Patchett AA, Smith RG. Identification of a new G-protein-linked receptor for growth hormone secretagogues. Mol Endocrinol. 1996; 10:57-61.

25. Muccioli G, Ghè C, Ghigo MC, et al. Specific receptors for synthetic GH secretagogues in the human brain and pituitary gland. J Endocrinol. 1998; 157:99-106.

26. Papotti M, Ghè C, Cassoni P, et al. Growth hormone secretagogue binding sites in peripheral human tissues. J Clin Endocrinol Metab. 2000; 85:3803-7.

27. Cassoni P, Papotti M, Ghè C, et al. Identification, characterization and biological activity of specific receptors for natural (ghrelin) and synthetic growth hormone secretagogues in human breast carcinomas and cell lines. J Clin Endocrinol Metab. 2001; 86:1738-45.

28. Ghè C, Cassoni P, Catapano F, et al. The antiproliferative effect of synthetic peptidyl GH secretagogues in human CALU-1 lung carcinoma cells. Endocrinology. 2002; 143:484-91.

29. Muccioli G, Papotti M, Ong H, et al. Presence of specific receptors for synthetic growth hormone secretagogues in the human heart. Arch Pharm. 1998; 358:R549-.

30. Ong H, McNicoll N, Escher E, et al. Identification of a pituitary growth hormone-releasing peptide (GHRP) receptor subtype by photoaffinity labeling. Endocrinology. 1998; 139:432-5.

31. Bodart V, Febbraio M, Demers A, et al. CD36 mediates the cardiovascular action of growth hormone-releasing peptides in the heart. Circ Res. 2002; 90:844-9.

32. Bennett PA, Thomas GB, Howard AD, et al. Hypothalamic growth hormone secretagogue-receptor (GHS-R) expression is regulated by growth hormone in the rat. Endocrinology. 1997; 138:4552-7.

33. Guan XM, Yu H, Palyha OC, et al. Distribution of mRNA encoding the growth hormone secretagogue receptor in brain and peripheral tissues. Brain Res Mol Brain Res. 1997; 48:23-9.

34. Tannenbaum GS, Lapointe M, Beaudet A, Howard AD. Expression of growth hormone secretagogue-receptors by growth hormone-releasing hormone neurons in the mediobasal hypothalamus. Endocrinology. 1998; 139:4420-3.

35. Willesen MG, Kristensen P, Romer J. Co-localization of growth hormone secretagogue receptor and NPY mRNA in the arcuate nucleus of the rat. Neuroendocrinology. 1999; 70:306-16.

36. Mitchell V, Bouret S, Beauvillain JC, et al. Comparative distribution of mRNA encoding the growth hormone secretagogue-receptor (GHS-R) in Microcebus murinus (Primate, lemurian) and rat forebrain and pituitary. Comp Neurol. 2001; 429:469-89.

37. Barlier A, Zamora AJ, Grino M, et al. Expression of functional growth hormone secretagogue receptors in human pituitary adenomas: polymerase chain reaction, triple in-situ hybridization and cell culture studies. J Neuroendocrinol. 1999; 11:491-502.

38. Shuto Y, Shibasaki T, Wada K, et al. Generation of polyclonal antiserum against the growth hormone secretagogue receptor (GHS-R): evidence that the GHS-R exists in the hypothalamus, pituitary and stomach of rats. Life Sci. 2001; 68:991-6.

39. Adams EF, Huang B, Buchfelder M, et al. Presence of growth hormone secretagogue receptor messenger ribonucleic acid in human pituitary tumors and rat GH3 cells. J Clin Endocrinol Metab. 1998; 83:638-42.

40. Skinner MM, Nass R, Lopes B, Laws ER, Thorner MO. Growth hormone secretagogue receptor expression in human pituitary tumors. J Clin Endocrinol Metab. 1998; 83:4314-20.

41. Korbonits M, Jacobs RA, Aylwin SJ, et al. Expression of the growth hormone secretagogue receptor in pituitary adenomas and other neuroendocrine tumors. J Clin Endocrinol Metab. 1998; 83:3624-30.

42. de Keyzer Y, Lenne F, Bertagna X. Widespread transcription of the growth hormone-releasing peptide receptor gene in neuroendocrine human tumors. Eur J Endocrinol. 1997; 137:715-8.

43. Murata M, Okimura Y, Iida K, et al. Ghrelin modulates the downstream molecules of insulin signaling in hepatoma cells. J Biol Chem. 2002; 277:5667-74.

44. Baldanzi G, Filigheddu N, Cutrupi S, et al. Ghrelin and des-acyl ghrelin inhibit cell death in cardiomyocytes and endothelial cells through ERK1/2 and PI 3-kinase/AKT. J Cell Biol. 2002; 159:1029-37.

45. Deghenghi R. Diazaspiro, azepino and azabicyclo therapeutic peptides. United States Patent, Patent No. c6,025,471; 2000.

46. Muccioli G, Tschop M, Papotti M, Deghenghi R, Heiman M, Ghigo E. Neuroendocrine and peripheral activities of ghrelin: implications in metabolism and obesity. Eur J Pharmacol. 2002; 440:235-54.

47. Broglio F, Arvat E, Benso A, et al.. Ghrelin: much more than a natural growth hormone secretagogue. Isr Med Assoc J. 2002; 4:607-13.

Chapter 11

METABOLIC ACTIONS OF GHRELIN

Carlotta Gauna & Aart Jan van der Lely
Department of Internal Medicine, Section of Endocrinology, Erasmus MC, Rotterdam, The Netherlands

Abstract: Ghrelin is mainly expressed in the stomach, followed by lower parts of the gastrointestinal tract. Ghrelin secretion is mainly influenced by changes in energy balance and glucose homeostasis, followed by alterations of endocrine axes. Ghrelin therefore seems to be an interface between energy homeostasis, glucose metabolism and physiological processes regulated by the classical endocrine axes that e.g., control growth. Ghrelin most likely defends fat mass to ensure survival and to provide the calories that are needed for growth hormone to act on growth and repair. The possibility of using a ghrelin antagonist for the treatment of obesity initiated a hunt for such a valuable agent. It will not be long before antagonists are disclosed and tested for the treatment of obesity.

Key words: body weight, food intake, stomach, insulin, metabolism

1.   INTRODUCTION

Apart from a potent growth hormone (GH)-releasing action, ghrelin has many other actions including stimulation of the lactotroph and corticotroph, influence on the pituitary gonadal axis, stimulation of appetite, control of energy balance, influence on sleep and behavior, control of gastric motility and acid secretion, influence on pancreatic exocrine and endocrine function as well as on glucose metabolism. Moreover, many growth hormone secretagogues (GHS) and/or ghrelin have cardiovascular actions and modulate proliferation of neoplastic cells, as well as of the immune system. In this chapter we will discuss the metabolic actions of this fascinating new hormone.

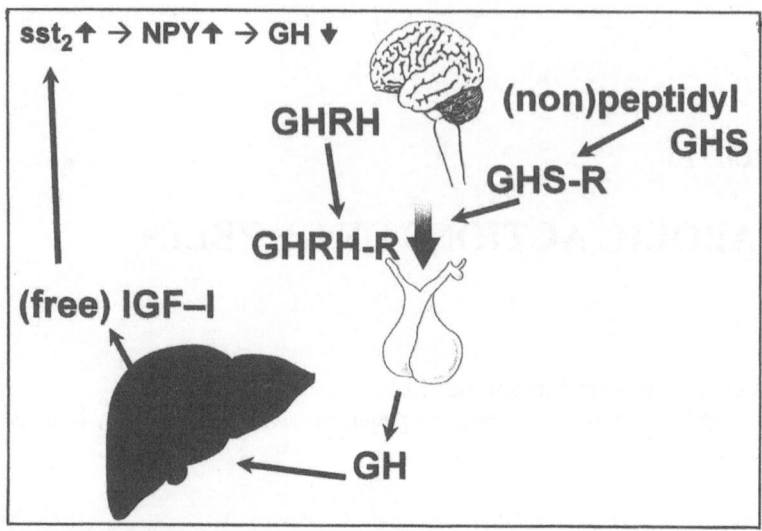

Figure 11-1. The collaboration between the growth hormone releasing hormone (GHRH) and the growth hormone secretagogue (GHS) receptor (GHS-R) in the increase in growth hormone (GH) secretion. sst_2: somatostatin receptor type 2; NPY: neuropeptide Y; IGF-I: insulin-like growth factor-I.

2.   PERIPHERAL ACTIONS OF SYNTHETIC AND NATURAL GHS

The first synthesized GHS were non-natural peptides, which were designed rather than isolated by Bowers and Momany in the late 1970s. They were met-enkephalin derivatives devoid of any opioid activity (1,2). Growth Hormone Releasing Peptide-6 was the first hexapeptide to actively release GH *in vivo*. A well known member of the non-peptidyl GHS family is spiroindoline L-163,191 (MK-0677) (3-11). MK-0677 has been shown to possess a high bioavailability and it is able to enhance 24-hour GH secretion after single oral administration (3-11). Studies focusing on the distribution of the identified GHS receptors (GHS-R) showed a particular concentration of these receptors in the hypothalamus-pituitary area, but specific binding sites have also been found in other brain areas and peripheral, endocrine and nonendocrine animal and human tissues (12-17). Indeed this distribution of the GHS-R does explain the GH-releasing effect of GHS, but also their other endocrine and nonendocrine biological activities (Figure 11-1 and 11-2) (3;18-31).

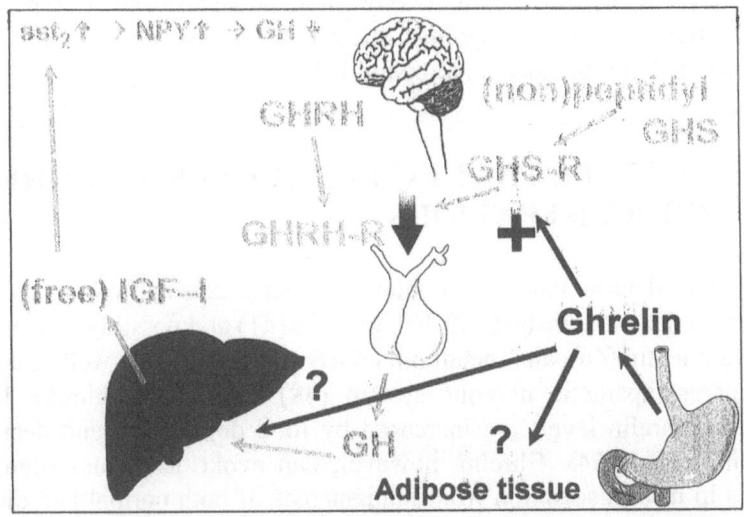

Figure 11-2. The central effects of ghrelin on growth hormone (GH) secretion are likely accompanied by direct, GH independent effects on several peripheral systems, involved in metabolic control (e.g., liver, adipose tissue). GHRH: growth hormone releasing hormone; GHRH-R: GHRH receptor; GHS: growth hormone secretagogue, GHS-R: GHS receptor; GH: growth hormone; sst$_2$: somatostatin receptor type 2; NPY: neuropeptide Y; IGF-I: insulin-like growth factor-I

While the majority of circulating ghrelin is produced by the stomach and the small bowel (15,32), ghrelin is expressed in a variety of tissues which include the stomach, the intestine, the pituitary, the placenta, lymphocytes, the testes, the lungs, the kidney, the pancreas and the hypothalamus (14-17,23,31,33-38).

While a classical endocrine role for ghrelin as a peptide hormone seems evident, paracrine activities of ghrelin might play an additional role (15,32). Removal of the stomach or the acid-producing part of the stomach in rats reduces serum ghrelin concentration by ~ 80%, further supporting the view that the stomach is the main source of this endogenous GHS-R ligand (15,32).

Korbonits and coworkers have detected small amounts of ghrelin in adrenal glands, esophagus, adipocytes, gall bladder, muscle, myocardium, ovary, prostate, skin, spleen, thyroid, blood vessels and liver, using real time reverse transcriptase-polymerase chain reaction (13). In addition, prepro-ghrelin production is shown in rat mesangial cells and mouse podocytes, indicating the production of ghrelin in kidney, glomerulus and renal cells and suggesting possible paracrine roles of ghrelin in the kidney (33).

In summary, ghrelin is mainly expressed in the stomach, followed by lower parts of the gastrointestinal tract. Ghrelin expression levels in other organs are relatively low. While its physiological significance as a paracrine factor is the subject of ongoing studies, an endocrine role for extra-gastrointestinal ghrelin appears to be unlikely.

3.   FACTORS THAT INFLUENCE SERUM GHRELIN CONCENTRATIONS

Only few determinants of circulating ghrelin concentration have been identified to date, i.e., insulin (39,40), glucose (41) and possibly GH (42-45). Also serum leptin (46), and melatonin concentrations (47), as well as activity of the parasympathetic nervous system (48) can modify ghrelin levels. Circulating ghrelin levels are increased by food deprivation and decreased by feeding (19,49-54). Ghrelin, however, can evoke large and significant increases in insulin secretion from the pancreas of both normal and diabetic rats (55). Ingestion of sugar suppresses ghrelin secretion in rats (19). These observations indicate a direct inhibitory effect of food intake on ghrelin containing X/A-like cells in the oxyntic mucosa of the rat stomach rather than an exclusively insulin-mediated effect. That insulin is an independent determinant of the circulating ghrelin concentration has recently been shown by several study groups using hyperinsulinemic euglycemic clamp studies in humans (39,40). Ghrelin expression can be stimulated in rats by insulin-induced hypoglycemia, leptin administration and central leptin gene therapy (46,51). However, unlike food intake, the administration of insulin and glucose does not suppress ghrelin levels. These data suggest that the suppressive effect of food intake or oral glucose on serum ghrelin is not likely to be mediated by the changes of plasma insulin and glucose observed after the ingestion (56).

Studies by Cummings and coworkers (53,57) indicate that each daily meal is followed by decreases of circulating ghrelin levels, while ghrelin administration in healthy volunteers causes hunger sensations (21,58,59).

In summary, ghrelin expression as well as ghrelin secretion are mainly influenced by changes in energy balance and glucose homeostasis, followed by alterations of endocrine axes. Ghrelin therefore seems to be an interface between energy homeostasis, glucose metabolism and physiological processes regulated by the classical endocrine axes that control growth.

4.   GASTRO-ENTERO-PANCREATIC ACTIONS OF GHRELIN RECEPTOR AGONISTS

It is not surprising that the gastric hormone ghrelin can act at the gastro-entero-pancreatic level as well, where GHS-R type 1a and 1b expression have been demonstrated (13,15,35,51,60-64).

Ghrelin stimulates gastric acid secretion and motility in rats (48,65,66) and circulating ghrelin levels are correlated with gastric emptying time in humans (49). These actions are mediated by the cholinergic system as they can be abolished by muscarinic blockade (48), while this acetylcholine-mediated stimulatory effect of ghrelin on gastric acid secretion is partially mediated by the central nervous system (67).

Ghrelin has been demonstrated to be expressed by pancreatic endocrine α-cells in rat and human tissue by some authors (68) and by pancreatic β-cells by one group only (69). Animal studies show conflicting results regarding the influence of ghrelin on insulin secretion (24,68,70). In fact, ghrelin was able to stimulate insulin secretion from isolated rat pancreatic islets (68) and in rats *in vivo* as well (55,70). On the other hand, insulin secretion from isolated rat pancreas, perfused *in situ* after stimulation with glucose, arginine, and carbachol, was found to be blunted by exposure to ghrelin that also reduced the somatostatin response to arginine (24). These findings suggest that ghrelin exerts a tonic inhibitory regulation on insulin secretion from pancreatic β-cells, contributing at least in rats to a restrained release during food deprivation. In agreement with this hypothesis, a clear negative association between ghrelin and insulin secretion has been found in humans as well as in animals by the majority of authors (39,51,57,71-73), although not by others (56). In humans, ghrelin induces a significant increase in human plasma glucose levels which is surprisingly followed by a reduction in insulin secretion (21). Coupled with the observation that acute, as well as chronic treatment with GHS, particularly non-peptidyl derivatives, induced hyperglycemia and insulin resistance in a considerable number of elderly subjects and obese patients (7,74,75). These observations suggest that ghrelin is a gastro-entero-pancreatic hormone, exerting a significant role in the fine-tuning of insulin secretion and that in the meantime also affects glucose metabolism.

Ghrelin blocks the inhibitory effects of insulin on gluconeogenesis (76). However, it is also suggested that ghrelin could have direct stimulatory effects on glycogenolysis, and this action is likely mediated by a non-GHS-R type 1a process, as it is not exerted by synthetic GHS (21). All data show that ghrelin secretion seems to be negatively associated with body mass while its levels are increased in anorexia nervosa and cachexia. Also early morning, overnight fasting ghrelin concentrations seem to be decreased in obese subjects. This indicates the potential major impact of ghrelin on insulin secretion and glucose metabolism suggested by studies summarized above (39,41,45,75-78). The only notable exception, in this context, to the negative association between ghrelin and body mass is represented by obese patients with Prader-Willi syndrome who show peculiarly elevated circulating ghrelin levels (79).

5.   GHRELIN AGONISTS AND THE REGULATION OF METABOLISM

GHS (including ghrelin) have direct and non-GH dependent actions on metabolism. In normal individuals, preprandial GHS administration significantly increases postprandial glucose and insulin levels, but only in the presence of the GH receptor antagonist pegvisomant, i.e., in an acute model of GH deficiency. This hyperinsulinism is accompanied by a rapid decrease of free fatty acids. These GHS-mediated changes indicate that when GH bioactivity is low, GHS can induce changes in metabolic control, characteristic for "the metabolic syndrome". Because in this study GH-action was blocked by pegvisomant, these GHS-mediated metabolic changes in the "gastro-entero-hepatic axis" must be direct and non-GH/non-pituitary mediated (75).

This study clearly shows that fasting induces an acute and distinct diurnal rhythm in systemic ghrelin concentrations that is not present in the fed state. These changes in serum ghrelin levels during fasting are followed by similar changes in serum GH concentrations, indicating that ghrelin is the driving force of increased GH secretion during fasting (45). This phenomenon could not be explained by changes in insulin, glucose or free fatty acid levels.

Apart from these effects of GHS on metabolism which take hours, indicating changes in gene and/or protein expression, direct effects of ghrelin on glucose and insulin have also been demonstrated. In humans, ghrelin has distinct and immediate effects on important determinants of metabolism, i.e., ghrelin administration immediately increases glucose, while it lowers insulin and glucagon (21). Moreover, daily ghrelin administration in rodents for only several days induces an obese state, again indicating that these ghrelin-mediated effects on metabolism are powerful and clinically relevant (19). These data indicate that GHS-mediated effects are involved in the induction of the metabolic alterations, as well as subsequent changes in body composition, which are characteristic for the insulin resistance syndrome (metabolic syndrome), as observed in GH deficiency, but also during normal aging. Also, the presence of ghrelin could be detected in human carcinoid tumors of the foregut, as well as in bowel, pancreas, kidneys, the immune system, placenta, testes, pituitary, and hypothalamus. Although specific binding of ghrelin in many tissues could be detected, frequently no ghrelin receptor (GHS-R type 1a) expression could be found, however.

Ghrelin and/or ghrelin action are very well preserved throughout many species indicating an important biological role for this peptide hormone. Ghrelin expression and/or its biological action have been shown in creatures as diverse as fish, birds, amphibians and mammals (52,80-91). Ghrelin appears to ensure that sufficient amounts of energy are available for GH to act on growth and repair (92), and ghrelin signals the brain when energy has to be stored or saved (93).

Several recent studies show that circulating ghrelin peptide levels are decreased in obese individuals (53,94-96), while ghrelin levels are increased in cachexia and in anorectic conditions (71,97-99), as well as during food deprivation (19,51,71). However, a further reduction of the already low plasma ghrelin concentrations in obese individuals can still trigger the reduction of body fat mass or at least prevent recidivism of obesity following diet-induced weight loss, as has recently been shown by Cummings and coworkers (53). Following gastric bypass surgery, obese patients exhibited a more impressive weight loss than can generally be achieved by voluntary caloric restriction. While these adaptive changes in ghrelin secretion might contribute to the spreading obesity epidemic, only the availability of a potent ghrelin antagonist will allow us to test if pharmacological "simulation" of a "gastric bypass" could represent an effective treatment option for obesity.

Leptin might be the signal from adipocytes that informs ghrelin secreting α-cells in the gastrointestinal tract on the extent of stored fat, since ghrelin levels in these obese rats are high as compared to weight matched controls, indicating that, at least in part, the activation of the leptin receptor at the levels of the ghrelin secreting cell has a suppressing influence (54). It seems however possible that in case of a constant exposure of ghrelin receptors to high concentrations of ghrelin receptor ligand, a down-regulation of receptor sensitivity can be seen, similar to the leptin resistance syndrome in hyperleptinemic obese patients (100). If this scenario turns out to be true, it might be difficult to establish ghrelin or one of its receptor agonists as a treatment for anorexia or cachexia, although high doses might still show a favorable effect.

In summary, ghrelin most likely defends fat mass to ensure survival and to provide the calories that are needed for GH to act on growth and repair. The strong orexigenic and adipogenic drive of ghrelin, however, can produce an overshoot into the obese range in the presence of a highly palatable and abundant diet (101).

6.　POTENTIAL CLINICAL IMPLICATIONS OF GHRELIN IN THE MODIFICATION OF METABOLISM

Until ghrelin was exposed as a hormone that stimulates appetite in rodents and humans there was no need for a ghrelin antagonist. Although such a molecule may be expected to reduce GH levels and thus could be indicated for the treatment of acromegaly, other effective agents are available. The new possibility of using a ghrelin antagonist for the treatment of obesity initiated a hunt for such a valuable agent. Thus, it will not be long before antagonists are disclosed and tested for the treatment of obesity.

REFERENCES

1. Bowers CY. Growth hormone-releasing peptide (GHRP). Cell Mol Life Sci. 1998; 54:1316-29.
2. Momany FA, Bowers CY, Reynolds GA, Chang D, Hong A, Newlander K. Design, synthesis, and biological activity of peptides which release growth hormone *in vitro*. Endocrinology. 1981; 108:31-9.
3. Smith RG, Van der Ploeg LH, Howard AD, et al. Peptidomimetic regulation of growth hormone secretion. Endocr Rev. 1997; 18:621-45.
4. Patchett AA, Nargund RP, Tata JR, et al. Design and biological activities of L-163,191 (MK-0677): a potent, orally active growth hormone secretagogue. Proc Natl Acad Sci U S A. 1995; 92:7001-5.
5. Chapman IM, Bach MA, Van Cauter E, et al. Stimulation of the growth hormone (GH)-insulin-like growth factor I axis by daily oral administration of a GH secretogogue (MK-677) in healthy elderly subjects. J Clin Endocrinol Metab. 1996; 81:4249-57.
6. Thorner MO. Theodore R. Woodward Award. Age-related decline in growth hormone secretion: clinical significance and potential reversibility. Trans Am Clin Climatol Assoc. 1996; 108:99-105; discussion 105-8.
7. Chapman IM, Pescovitz OH, Murphy G, et al. Oral administration of growth hormone (GH) releasing peptide-mimetic MK-677 stimulates the GH/insulin-like growth factor-I axis in selected GH-deficient adults. J Clin Endocrinol Metab. 1997; 82:3455-63.
8. Svensson J, Ohlsson C, Jansson JO, et al. Treatment with the oral growth hormone secretagogue MK-677 increases markers of bone formation and bone resorption in obese young males. J Bone Miner Res. 1998; 13:1158-66.
9. Schleim KD, Jacks T, Cunningham P, et al. Increases in circulating insulin-like growth factor I levels by the oral growth hormone secretagogue MK-0677 in the beagle are dependent upon pituitary mediation. Endocrinology. 1999; 140:1552-8.
10. Murphy MG, Weiss S, McClung M, et al. Effect of alendronate and MK-677 (a growth hormone secretagogue), individually and in combination, on markers of bone turnover and bone mineral density in postmenopausal osteoporotic women. J Clin Endocrinol Metab. 2001; 86:1116-25.

11. Gertz BJ, Barrett JS, Eisenhandler R, et al. Growth hormone response in man to L-692,429, a novel nonpeptide mimic of growth hormone-releasing peptide-6. J Clin Endocrinol Metab. 1993; 77:1393-7.
12. Bluet-Pajot MT, Tolle V, Zizzari P, et al. Growth hormone secretagogues and hypothalamic networks. Endocrine. 2001; 14:1-8.
13. Gnanapavan S, Kola B, Bustin SA, et al. The tissue distribution of the mRNA of ghrelin and subtypes of its receptor, GHS-R, in humans. J Clin Endocrinol Metab. 2002; 87:2988-91.
14. Kojima M, Hosoda H, Date Y, Nakazato M, Matsuo H, Kangawa K. Ghrelin is a growth-hormone-releasing acylated peptide from stomach. Nature. 1999; 402:656-60.
15. Date Y, Kojima M, Hosoda H, et al. Ghrelin, a novel growth hormone-releasing acylated peptide, is synthesized in a distinct endocrine cell type in the gastrointestinal tracts of rats and humans. Endocrinology. 2000; 141:4255-61.
16. Korbonits M, Kojima M, Kangawa K, Grossman AB. Presence of ghrelin in normal and adenomatous human pituitary. Endocrine. 2001; 14:101-4.
17. Muccioli G, Tschop M, Papotti M, Deghenghi R, Heiman M, Ghigo E. Neuroendocrine and peripheral activities of ghrelin: implications in metabolism and obesity. Eur J Pharmacol. 2002; 440:235-54.
18. Cassoni P, Papotti M, Ghe C, et al. Identification, characterization, and biological activity of specific receptors for natural (ghrelin) and synthetic growth hormone secretagogues and analogs in human breast carcinomas and cell lines. J Clin Endocrinol Metab. 2001; 86:1738-45.
19. Tschop M, Smiley DL, Heiman ML. Ghrelin induces adiposity in rodents. Nature. 2000; 407:908-13.
20. Wren AM, Small CJ, Ward HL, et al. The novel hypothalamic peptide ghrelin stimulates food intake and growth hormone secretion. Endocrinology. 2000; 141:4325-8.
21. Broglio F, Arvat E, Benso A, et al. Ghrelin, a natural GH secretagogue produced by the stomach, induces hyperglycemia and reduces insulin secretion in humans. J Clin Endocrinol Metab. 2001; 86:5083-6.
22. Wren AM, Seal LJ, Cohen MA, et al. Ghrelin enhances appetite and increases food intake in humans. J Clin Endocrinol Metab. 2001; 86:5992-5.
23. Tena-Sempere M, Barreiro ML, Gonzalez LC, et al. Novel expression and functional role of ghrelin in rat testis. Endocrinology. 2002; 143:717-25.
24. Egido EM, Rodriguez-Gallardo J, Silvestre RA, Marco J. Inhibitory effect of ghrelin on insulin and pancreatic somatostatin secretion. Eur J Endocrinol. 2002; 146:241-4.
25. Zhang W, Chen M, Chen X, Segura BJ, Mulholland MW. Inhibition of pancreatic protein secretion by ghrelin in the rat. J Physiol. 2001; 537:231-6.
26. Jeffery PL, Herington AC, Chopin LK. Expression and action of the growth hormone releasing peptide ghrelin and its receptor in prostate cancer cell lines. J Endocrinol. 2002; 172:R7-11.
27. Casanueva FF, Dieguez C. Growth Hormone Secretagogues: Physiological Role and Clinical Utility. Trends Endocrinol Metab. 1999; 10:30-8.
28. Bodart V, Febbraio M, Demers A, et al. CD36 mediates the cardiovascular action of growth hormone-releasing peptides in the heart. Circ Res. 2002; 90:844-9.

29. Okumura H, Nagaya N, Enomoto M, Nakagawa E, Oya H, Kangawa K. Vasodilatory effect of ghrelin, an endogenous peptide from the stomach. J Cardiovasc Pharmacol. 2002; 39:779-83.

30. Yoshihara F, Kojima M, Hosoda H, Nakazato M, Kangawa K. Ghrelin: a novel peptide for growth hormone release and feeding regulation. Curr Opin Clin Nutr Metab Care. 2002; 5:391-5.

31. Hattori N, Saito T, Yagyu T, Jiang BH, Kitagawa K, Inagaki C. GH, GH receptor, GH secretagogue receptor, and ghrelin expression in human T cells, B cells, and neutrophils. J Clin Endocrinol Metab. 2001; 86:4284-91.

32. Dornonville de la Cour, Bjorkqvist M, Sandvik AK, et al. A-like cells in the rat stomach contain ghrelin and do not operate under gastrin control. Regul Pept. 2001; 99:141-50.

33. Mori K, Yoshimoto A, Takaya K, et al. Kidney produces a novel acylated peptide, ghrelin. FEBS Lett. 2000; 486:213-6.

34. Gualillo O, Caminos J, Blanco M, et al. Ghrelin, a novel placental-derived hormone. Endocrinology. 2001; 142:788-94.

35. Volante M, Papotti M, Gugliotta P, Migheli A, Bussolati G. Extensive DNA fragmentation in oxyphilic cell lesions of the thyroid. J Histochem Cytochem. 2001; 49:1003-11.

36. Tanaka M, Hayashida Y, Nakao N, Nakai N, Nakashima K. Testis-specific and developmentally induced expression of a ghrelin gene-derived transcript that encodes a novel polypeptide in the mouse. Biochim Biophys Acta. 2001; 1522:62-5.

37. Chapman IM, Hartman ML, Pezzoli SS, Thorner MO. Enhancement of pulsatile growth hormone secretion by continuous infusion of a growth hormone-releasing peptide mimetic, L-692,429, in older adults - a clinical research center study. J Clin Endocrinol Metab. 1996; 81:2874-80.

38. Volante M, Fulcheri E, Allìa E, Cerrato M, Pucci A, Papotti M. Ghrelin expression in fetal, infant, and adult human lung. J Histochem Cytochem. 2002; 50:1013-21.

39. Saad MF, Bernaba B, Hwu CM, et al. Insulin regulates plasma ghrelin concentration. J Clin Endocrinol Metab. 2002; 87:3997-4000.

40. Mohlig M, Spranger J, Otto B, Ristow M, Tschop M, Pfeiffer AF. Euglycemic hyperinsulinemia, but not lipid infusion, decreases circulating ghrelin levels in humans. J Endocrinol Invest 2002; 25:RC36-8.

41. Nakagawa E, Nagaya N, Okumura H, et al. Hyperglycaemia suppresses the secretion of ghrelin, a novel growth-hormone-releasing peptide: responses to the intravenous and oral administration of glucose. Clin Sci (Lond). 2002; 103:325-8.

42. Tschop M, Flora DB, Mayer JP, Heiman ML. Hypophysectomy prevents ghrelin-induced adiposity and increases gastric ghrelin secretion in rats. Obes Res. 2002; 10:991-9.

43. Cappiello V, Ronchi C, Morpurgo PS, et al. Circulating ghrelin levels in basal conditions and during glucose tolerance test in acromegalic patients. Eur J Endocrinol. 2002; 147:189-94.

44. van der Toorn FM, Janssen JA, De Herder WW, Broglio F, Ghigo E, Van Der Lely AJ. Central ghrelin production does not substantially contribute to systemic ghrelin concentrations; a study in two subjects with active acromegaly. Eur J Endocrinol. 2002; 147:195-9.

45. Muller AF, Lamberts SW, Janssen JA, et al. Ghrelin drives GH secretion during fasting in man. Eur J Endocrinol. 2002; 146:203-7.
46. Kalra SP, Dube MG, Pu S, Xu B, Horvath TL, Kalra PS. Interacting appetite-regulating pathways in the hypothalamic regulation of body weight. Endocr Rev. 1999; 20:68-100.
47. Mustonen AM, Nieminen P, Hyvarinen H. Preliminary evidence that pharmacologic melatonin treatment decreases rat ghrelin levels. Endocrine. 2001; 16:43-6.
48. Masuda Y, Tanaka T, Inomata N, et al. Ghrelin stimulates gastric acid secretion and motility in rats. Biochem Biophys Res Commun. 2000; 276:905-8.
49. Tschop M, Wawarta R, Riepl RL, et al. Post-prandial decrease of circulating human ghrelin levels. J Endocrinol Invest. 2001; 24:RC19-21.
50. Asakawa A, Inui A, Kaga T, et al. Ghrelin is an appetite-stimulatory signal from stomach with structural resemblance to motilin. Gastroenterology. 2001; 120:337-45.
51. Toshinai K, Mondal MS, Nakazato M, et al. Upregulation of Ghrelin expression in the stomach upon fasting, insulin-induced hypoglycemia, and leptin administration. Biochem Biophys Res Commun. 2001; 281:1220-5.
52. Hayashida T, Murakami K, Mogi K, et al. Ghrelin in domestic animals: distribution in stomach and its possible role. Domest Anim Endocrinol. 2001; 21:17-24.
53. Cummings DE, Weigle DS, Frayo RS, et al. Plasma ghrelin levels after diet-induced weight loss or gastric bypass surgery. N Engl J Med. 2002; 346:1623-30.
54. Ariyasu H, Takaya K, Hosoda H, et al. Delayed Short-Term Secretory Regulation of Ghrelin in Obese Animals: Evidenced by a Specific RIA for the Active Form of Ghrelin. Endocrinology. 2002; 143:3341-50.
55. Adeghate E, Ponery AS. Ghrelin stimulates insulin secretion from the pancreas of normal and diabetic rats. J Neuroendocrinol. 2002; 14:555-60.
56. Caixas A, Bashore C, Nash W, Pi-Sunyer F, Laferrere B. Insulin, unlike food intake, does not suppress ghrelin in human subjects. J Clin Endocrinol Metab. 2002; 87:1902-6.
57. Cummings DE, Purnell JQ, Frayo RS, Schmidova K, Wisse BE, Weigle DS. A preprandial rise in plasma ghrelin levels suggests a role in meal initiation in humans. Diabetes. 2001; 50:1714-9.
58. Arvat E, Di Vito L, Broglio F, et al. Preliminary evidence that Ghrelin, the natural GH secretagogue (GHS)-receptor ligand, strongly stimulates GH secretion in humans. J Endocrinol Invest. 2000; 23:493-5.
59. Arvat E, Maccario M, Di Vito L, et al. Endocrine activities of ghrelin, a natural growth hormone secretagogue (GHS), in humans: comparison and interactions with hexarelin, a nonnatural peptidyl GHS, and GH-releasing hormone. J Clin Endocrinol Metab. 2001; 86:1169-74.
60. Guan XM, Yu H, Palyha OC, et al. Distribution of mRNA encoding the growth hormone secretagogue receptor in brain and peripheral tissues. Brain Res Mol Brain Res. 1997; 48:23-9.
61. Shuto Y, Shibasaki T, Wada K, et al. Generation of polyclonal antiserum against the growth hormone secretagogue receptor (GHS-R): evidence that the GHS-R exists in the hypothalamus, pituitary and stomach of rats. Life Sci. 2001; 68:991-6.
62. Papotti M, Cassoni P, Volante M, Deghenghi R, Muccioli G, Ghigo E. Ghrelin-producing endocrine tumors of the stomach and intestine. J Clin Endocrinol Metab. 2001; 86:5052-9.

63. Hosoda H, Kojima M, Matsuo H, Kangawa K. Purification and characterization of rat des-Gln14-Ghrelin, a second endogenous ligand for the growth hormone secretagogue receptor. J Biol Chem. 2000; 275:21995-22000.

64. Sakata I, Nakamura K, Yamazaki M, et al. Ghrelin-producing cells exist as two types of cells, closed- and opened-type cells, in the rat gastrointestinal tract. Peptides. 2002; 23:531-6.

65. Trudel L, Tomasetto C, Rio MC, et al. Ghrelin/motilin-related peptide is a potent prokinetic to reverse gastric postoperative ileus in rat. Am J Physiol Gastrointest Liver Physiol. 2002; 282:G948-52.

66. Murakami N, Hayashida T, Kuroiwa T, et al. Role for central ghrelin in food intake and secretion profile of stomach ghrelin in rats. J Endocrinol. 2002; 174:283-8.

67. Date Y, Nakazato M, Murakami N, Kojima M, Kangawa K, Matsukura S. Ghrelin acts in the central nervous system to stimulate gastric acid secretion. Biochem Biophys Res Commun. 2001; 280:904-7.

68. Date Y, Nakazato M, Hashiguchi S, et al. Ghrelin is present in pancreatic alpha-cells of humans and rats and stimulates insulin secretion. Diabetes. 2002; 51:124-9.

69. Volante M, AlIIa E, Gugliotta P, et al. Expression of ghrelin and of the GH secretagogue receptor by pancreatic islet cells and related endocrine tumors. J Clin Endocrinol Metab. 2002; 87:1300-8.

70. Lee HM, Wang G, Englander EW, Kojima M, Greeley GH Jr. Ghrelin, a new gastrointestinal endocrine peptide that stimulates insulin secretion: enteric distribution, ontogeny, influence of endocrine, and dietary manipulations. Endocrinology. 2002; 143:185-90.

71. Ariyasu H, Takaya K, Tagami T, et al. Stomach is a major source of circulating ghrelin, and feeding state determines plasma ghrelin-like immunoreactivity levels in humans. J Clin Endocrinol Metab. 2001; 86:4753-8.

72. Tolle V, Bassant MH, Zizzari P, et al. Ultradian rhythmicity of ghrelin secretion in relation with GH, feeding behavior, and sleep-wake patterns in rats. Endocrinology. 2002; 143:1353-61.

73. Pagotto U, Gambineri A, Vincennati V, Heiman ML, Tschop M, Pasquali R. Plasma ghrelin, obesity and the polycystic ovary syndrome: Correlation with insulin resistance and androgen levels. J Clin Endocrinol Metab. 2002; 87:5625-9.

74. Svensson J, Lonn L, Jansson JO, et al. Two-month treatment of obese subjects with the oral growth hormone (GH) secretagogue MK-677 increases GH secretion, fat-free mass, and energy expenditure. J Clin Endocrinol Metab. 1998; 83:362-9.

75. Muller AF, Janssen JA, Hofland LJ, et al. Blockade of the growth hormone (GH) receptor unmasks rapid GH-releasing peptide-6-mediated tissue-specific insulin resistance. J Clin Endocrinol Metab. 2001; 86:590-3.

76. Murata M, Okimura Y, Iida K, et al. Ghrelin modulates the downstream molecules of insulin signaling in hepatoma cells. J Biol Chem. 2002; 277:5667-74.

77. Cryer PE, Polonsky K. Glucose homeostasis and hypoglycemia. In: Wilson JD, ed. Williams Textbook of Endocrinology. Philadelphia: W.B. Saunders Company. 1998: 939-72.

78. Lucidi P, Murdolo G, Di Loreto C, et al. Ghrelin is not necessary for adequate hormonal counterregulation of insulin-induced hypoglycemia. Diabetes. 2002; 51:2911-4.

79. Cummings DE, Clement K, Purnell JQ, et al. Elevated plasma ghrelin levels in Prader Willi syndrome. Nat Med. 2002; 8:643-4.

80. Shepherd BS, Eckert SM, Parhar IS, et al. The hexapeptide KP-102 (D-ala-D-beta-Nal-ala-trp-D-phe-lys-NH(2)) stimulates growth hormone release in a cichlid fish (Ooreochromis mossambicus). J Endocrinol. 2000; 167:R7-10.

81. Furuse M, Tachibana T, Ohgushi A, Ando R, Yoshimatsu T, Denbow DM. Intracerebroventricular injection of ghrelin and growth hormone releasing factor inhibits food intake in neonatal chicks. Neurosci Lett. 2001; 301:123-6.

82. Kaiya H, Kojima M, Hosoda H, et al. Bullfrog ghrelin is modified by n-octanoic acid at its third threonine residue. J Biol Chem. 2001; 276:40441-8.

83. Nieminen P, Mustonen AM, Lindstrom-Seppa P, Asikainen J, Mussalo-Rauhamaa H, Kukkonen JV. Phytosterols act as endocrine and metabolic disruptors in the European polecat (Mustela putorius). Toxicol Appl Pharmacol. 2002; 178:22-8.

84. Nieminen P, Mustonen AM, Asikainen J, Hyvarinen H. Seasonal weight regulation of the raccoon dog (Nyctereutes procyonoides): interactions between melatonin, leptin, ghrelin, and growth hormone. J Biol Rhythms. 2002; 17:155-63.

85. Tomasetto C, Wendling C, Rio MC, Poitras P. Identification of cDNA encoding motilin related peptide/ghrelin precursor from dog fundus. Peptides. 2001; 22:2055-59.

86. Ahmed S, Harvey S. Ghrelin: a hypothalamic GH-releasing factor in domestic fowl (Gallus domesticus). J Endocrinol. 2002; 172:117-25.

87. Mustonen AM, Nieminen P, Hyvarinen H. Leptin, ghrelin, and energy metabolism of the spawning burbot (Lota lota, L.). J Exp Zool. 2002; 293:119-26.

88. Galas L, Chartrel N, Kojima M, Kangawa K, Vaudry H. Immunohistochemical localization and biochemical characterization of ghrelin in the brain and stomach of the frog Rana esculenta. J Comp Neurol. 2002; 450:34-44.

89. Mustonen AM, Nieminen P, Hyvarinen H. Melatonin and the Wintering Strategy of the Tundra Vole, Microtus oeconomus. Zoolog Sci. 2002; 19:683-7.

90. Sugino T, Hasegawa Y, Kikkawa Y, et al. A transient ghrelin surge occurs just before feeding in a scheduled meal-fed sheep. Biochem Biophys Res Commun. 2002; 295:255-60.

91. Riley LG, Hirano T, Grau EG. Rat Ghrelin Stimulates Growth Hormone and Prolactin Release in the Tilapia, Oreochromis mossambicus. Zoolog Sci. 2002; 19:797-800.

92. Heiman ML, Tschop M. Ghrelin provides the calories that growth hormone requires for growth and repair. Topic Endocrinol Suppl. 2001; 2:39-40.

93. Horvath TL, Diano S, Sotonyi P, Heiman M, Tschop M. Minireview: ghrelin and the regulation of energy balance--a hypothalamic perspective. Endocrinology. 2001; 142:4163-9.

94. Tschop M, Weyer C, Tataranni PA, Devanarayan V, Ravussin E, Heiman ML. Circulating ghrelin levels are decreased in human obesity. Diabetes. 2001; 50:707-9.

95. Hansen TK, Dall R, Hosoda H, et al. Weight loss increases circulating levels of ghrelin in human obesity. Clin Endocrinol. 2002; 56:203-6.

96. Shiiya T, Nakazato M, Mizuta M, et al. Plasma ghrelin levels in lean and obese humans and the effect of glucose on ghrelin secretion. J Clin Endocrinol Metab. 2002; 87:240-4.

97. Otto B, Cuntz U, Fruehauf E, et al. Weight gain decreases elevated plasma ghrelin concentrations of patients with anorexia nervosa. Eur J Endocrinol. 2001; 145:669-73.

98. Wisse BE, Frayo RS, Schwartz MW, Cummings DE. Reversal of cancer anorexia by blockade of central melanocortin receptors in rats. Endocrinology. 2001; 142:3292-301.

99. Nagaya N, Uematsu M, Kojima M, et al. Elevated circulating level of ghrelin in cachexia associated with chronic heart failure: relationships between ghrelin and anabolic/catabolic factors. Circulation. 2001; 104:2034-8.

100. Considine RV, Sinha MK, Heiman ML, et al. Serum immunoreactive-leptin concentrations in normal-weight and obese humans. N Engl J Med. 1996; 334:292-5.

101. Jequier E. Pathways to obesity. Int J Obes Relat Metab Disord. 2002; 26 (Suppl 2):S12-7.

Chapter 12

GASTRO-ENTERO-PANCREATIC ACTIONS OF GHRELIN

Yukari Date[1,2], Masamitsu Nakazato[1], Kenji Kangawa[2] & Hisayuki Matsuo[1]
[1]*Department of Internal Medicine, Miyazaki Medical College, Miyazaki, Japan,* [2]*National cardiovascular center research institute, Suita, Osaka, Japan*

Abstract: Ghrelin is primarily produced in the stomach and is known to be involved in feeding behavior, gastric function, and cardiovascular function, as well as growth hormone (GH) secretion. Ghrelin cells are believed to correspond morphologically to X/A-like cells whose products and functions have not yet been clarified. Identification of ghrelin cells in the stomach may contribute to the characterization of X/A-like cell function. Ghrelin is also present in the pancreatic islets and regulates insulin release, suggesting that ghrelin may play an important role in energy homeostasis. It is one of the major gut hormones and circulates through the bloodstream; however, ghrelin is thought to convey signals regulating GH release or feeding behavior and is transported to the brain via the vagal afferent system.

Key words: ghrelin cell, immunohistochemistry, gastric function, pancreatic ghrelin, vagal afferent system

1. GHRELIN IN THE GASTROINTESTINAL TRACT

1.1 Distribution of ghrelin cells

In 1999, ghrelin, a growth hormone (GH)-releasing acylated peptide, was isolated from rat and human stomachs (1), thus implicating this organ, and indeed the entire digestive system, in the process of GH secretion.

This finding came as a surprise to researchers who had been engaged in the discovery of ghrelin, because the concept that a gastrointestinal hormone could influence GH release from the pituitary was novel at time.

As early as gestational week 10, ghrelin cells are already detected in the developing gastrointestinal tract, pancreas, and lung (2). The number of ghrelin-immunoreactive cells in the rat fetal stomach increases as the stomach grows (3). The ghrelin content in the rat stomach also increases in an age-dependent manner from the neonate stage to the adult. Gastrectomy in rats decreases plasma ghrelin concentrations by approximately 80%, indicating that the stomach is the main source of circulating ghrelin (4,5). Ghrelin mRNA and ghrelin-immunoreactive cells are primarily located in the fundus of the stomach and in a scattered distribution from the neck to the base of the rat and human oxyntic gland (6). This characteristic pattern suggests that the ghrelin-producing cells comprise one of the endocrine cell populations of the stomach.

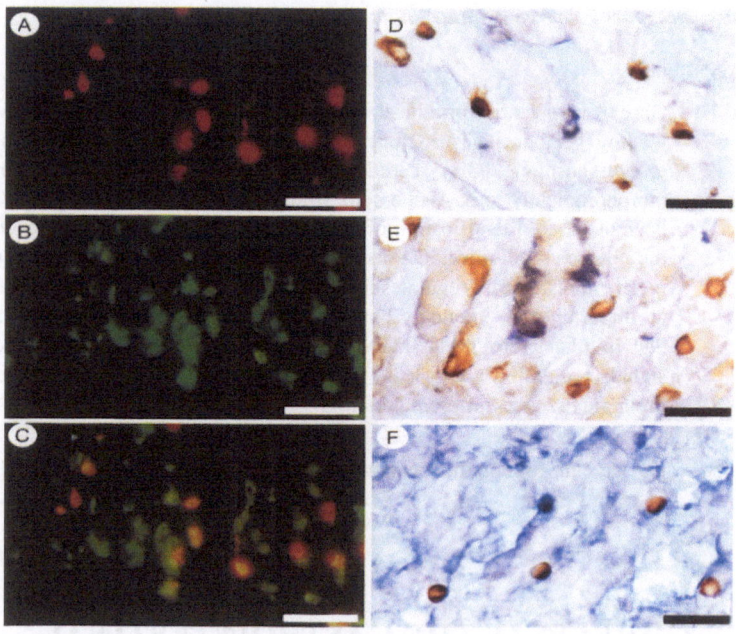

Figure 12-1. Localization of ghrelin-immunoreactive cells in the rat stomach. Antisera for ghrelin (A, C-F), chromogranin A (B), somatostatin (D), histidine decarboxylase (E), and serotonin (F) were used. A: ghrelin cells are visible in red by Alexa Flour 568 goat antirabbit IgG; B: chromogranin A cells are visible in green by Alex Flour 488 goat antimouse IgG; C: colocalization of ghrelin and chromogranin A cells shown in yellow; D: ghrelin cells are stained brown and somatostatin is stained blue; E: ghrelin cells are stained blue and histidine decarboxylase is stained brown; F: ghrelin cells are stained brown and serotonin is stained blue.

To evaluate this possibility, we investigated the colocalization between ghrelin and chromogranin A, a marker of endocrine cells. As expected, a double staining study showed that ghrelin cells accounted for about 20% of chromogranin A-immunoreactive endocrine cells and were clearly distinguished from somatostatin-producing D cells, histidine decarboxylase (HDC)-producing enterochromaffin-like (ECL) cells, and serotonin-producing enterochromaffin (EC) cells (Fig. 12-1).

The definitive characteristics of ghrelin cells were revealed by electron microscope immunohistochemistry to be round, compact, and electron dense ghrelin-containing granules (6) (Fig. 12-2).

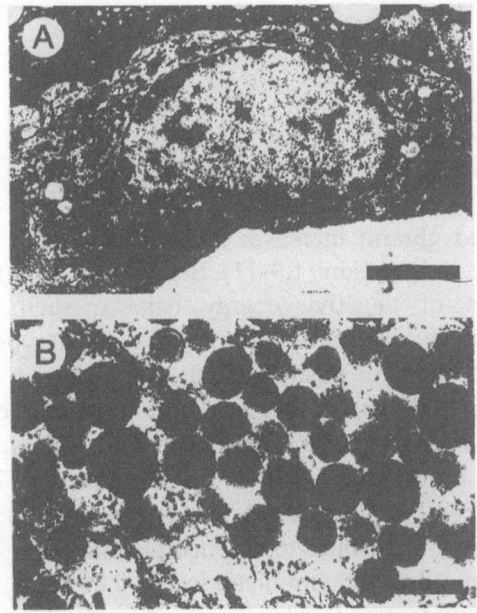

Figure 12-2. Immunoelectron photographs of a ghrelin cell in a rat oxyntic gland.

A: the ovoid cell has many round, compact, electron-dense granules in its cytoplasma; bar: 2 μm; magnification: ×7,000.

B: higher magnification of A, granules in the cytoplasm are labeled with immunogold staining for ghrelin; bar: 250 nm; magnification: ×36,000.

To date, four types of endocrine cells have been identified in the rat and human oxyntic mucosa through morphological criteria (7,8). The relative percentages of these four cells and their major products are shown in Table 12-1. Although X/A-like cells represent a major endocrine cell population in the oxyntic mucosa of both rats and humans, the composition of its intracellular granules has not been identified. As shown by immunohistochemical studies, X/A-like cells share many characteristics with ghrelin cells, including their localization, population, and ultrastructural features. Taken together, these results suggest that ghrelin cells should be considered to correspond to X/A-like cells and can thus be abbreviated as

"Gr cells", according to the standard nomenclature applied to other enteroendocrine cells.

Small numbers of ghrelin cells are also present in the upper small intestine (6), and even smaller numbers are present in the lower small intestine and large intestine.

Table 12-1. Relative percentage of endocrine cells in the oxyntic gland and their products

Cell type	Rat (%)	Human (%)	Products
ECL	60-70	30	Histamine
Gr (X/A-like)	20	20	Ghrelin
D	2-5	22	Somatostatin
EC	0-2	7	Serotonin
Others			

1.2   A role of ghrelin in gastric function

Intravenously (iv) administered ghrelin increases not only secretion of GH but also food intake and body weight gain (1,9-11). Secretion of ghrelin is upregulated under conditions of negative energy balance such as starvation, insulin-induced hypoglycemia, cachexia, and anorexia nervosa, and it is downregulated under conditions of positive energy balance such as feeding, hyperglycemia, and obesity (12-16). In this way, ghrelin would be a starvation signal produced in the stomach. Some peptides that influence feeding behavior, including corticotropin-releasing hormone (17), thyrotropin-releasing hormone (18,19), orexin A, and cocaine- and amphetamine-related transcript (20), contribute to the neural regulation of gastric acid secretion and gastric motility. Iv administration of 0.08 nmol to 2.0 nmol ghrelin to urethane-anesthetized rats increased gastric acid secretion and gastric motility in a dose dependent fashion (21). The maximum response in terms of gastric acid secretion was almost equally potent to that achieved with histamine (3.0 mg/kg, iv). These responses were abolished by pretreatment with either atropine (1.0 mg/kg, subcutaneously) or bilateral cervical vagotomy, but not by a histamine H_2-receptor antagonist (famotidine, 1.0 mg/kg, subcutaneously) (Fig. 12-3).

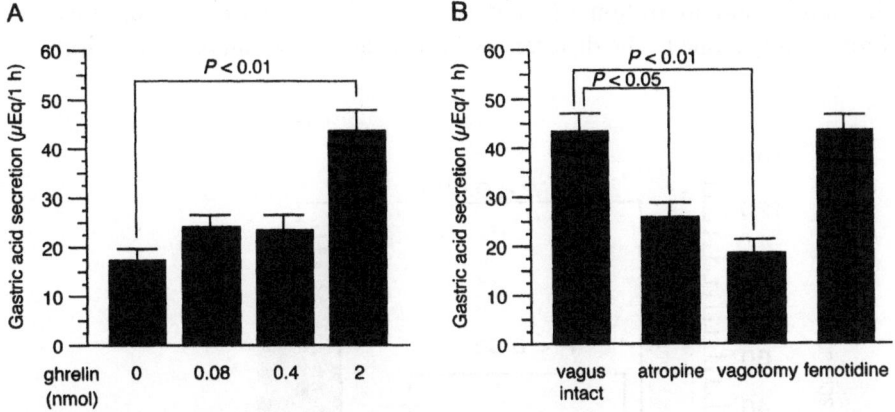

Figure 12-3. A: effect of iv administration of ghrelin on gastric acid output in urethane-anesthetized rats; B: effect of vagotomy, atropine, and femotidine on gastric acid output measured 1 hour after stimulation by iv administration of ghrelin (2 nmol) in urethane-anesthetized rats. Values are means ± standard error for 4 rats.

Although ghrelin is mainly produced in the stomach, it is also generated in the brain. Ghrelin-producing neurons are restricted to the hypothalamic arcuate nucleus, a region critical for feeding. In contrast, the ghrelin receptor is extensively distributed throughout the brain, including the hypothalamus and brainstem, areas that are central to the control of food intake and likely targets for substances affecting feeding behavior. Intracerebroventricularly (icv) administered ghrelin increases food intake and body weight gain as well as inducing GH release (22-24). We have already shown that ghrelin activates about 30% of neuropeptide Y (NPY)/agouti-related protein (AgRP) neurons. NPY and AgRP, which are co-located in neurons of the arcuate nucleus, have potent orexigenic effects. Furthermore, ghrelin-containing fibers are also known to project to the NPY/AgRP neurons. These findings indicate that central ghrelin plays a role in feeding behavior. Icv administration of ghrelin to urethane-anesthetized rats also increased gastric acid secretion in a dose-dependent manner, with a minimum active dose of a 1 nmol (Fig. 12-4) (25). Vagotomy or atropine administration abolished the gastric acid secretion induced by icv administration of ghrelin (Fig. 12-5). In addition, icv administration of ghrelin stimulated firing of gastric vagal efferent fibers. These two results together suggest that ghrelin stimulates gastric acid secretion via the vagus nerve. On the other hand, in conscious rats, icv administered ghrelin does not stimulate gastric acid secretion; rather, it inhibits it (26). This ghrelin-induced inhibition occurs when the icv administration occurs at a very low dose of under 1 nmol. The opposing stimulation and inhibition of acid secretion induced by icv administered ghrelin may be due to the differences in dose and experimental conditions.

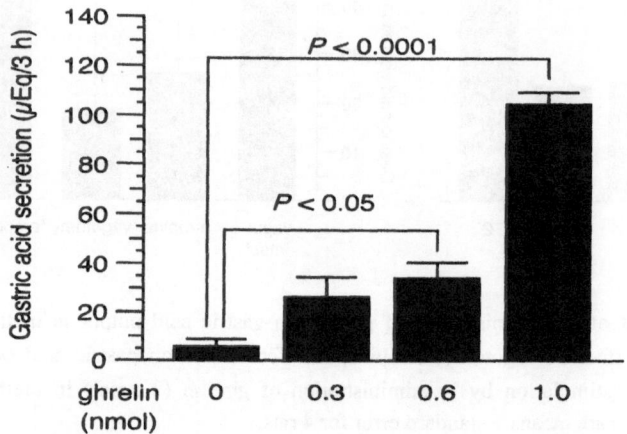

Figure 12-4. Effect of icv administration of ghrelin on gastric acid output in urethane-anesthetized rats. Values are means ± standard error for 5 rats.

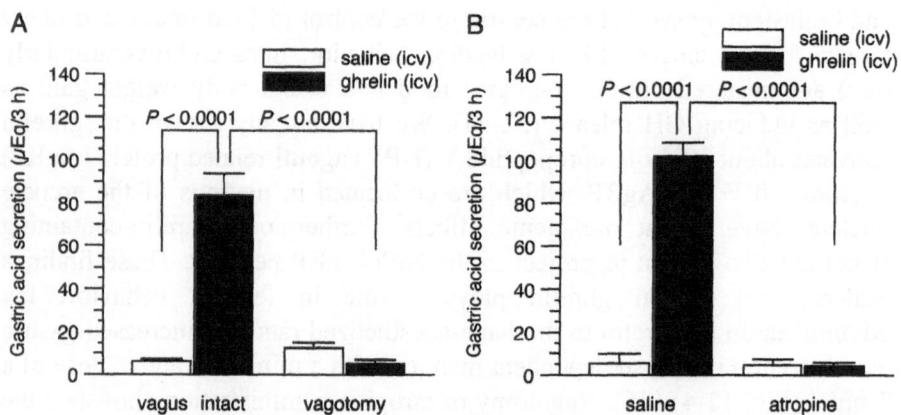

Figure 12-5. A: effect of vagotomy on gastric acid output measured 3 hours after stimulation by icv administration of ghrelin (1.0 nmol) in urethane-anesthetized rats; B: effect of atropine on 1.0 nmol icv ghrelin-induced gastric acid stimulation in urethane-anesthetized rats. Values are means ± standard error for 5 rats.

2.   GHRELIN AND GHRELIN RECEPTOR IN THE PANCREAS

2.1   Ghrelin and ghrelin receptor expression

The largest system of endocrine cells in the body is not the pituitary, the adrenals, or the thyroid; it is the enteroendocrine system–the endocrine cells of the digestive tract. The most familiar of these are the insulin- and glucagon-producing cells in the islets of the pancreas, but cell types closely related to these are present in every part of the gastrointestinal tract. More than 15 families of peptide hormones have been identified: they control such varied functions as glucose metabolism, delivery of bile and pancreatic juice into the gut lumen, renewal of the gut epithelium, and gut wall. Ghrelin is predominantly produced in the stomach; however, it is also expressed in other tissues including the pituitary (27), hypothalamus (1,28), heart, kidney (29), and pancreas (27,30-32). Endocrine cells of the gastrointestinal tract and the pancreas derive from multipotent endocrine stem cells. In the most highly evolved vertebrates, a close functional connection persists between the gastrointestinal endocrine cells and the pancreatic islets via the entero-insular axis (33). In humans, pancreatic ghrelin cells are numerous from midgestation to the early postnatal period, but decrease in number in adults (34). Ghrelin is also expressed in the rat pancreas with the small amount of this peptide (Fig. 12-6). The cellular source of ghrelin in the pancreas is still controversial. One group showed that ghrelin is present in pancreatic α-cells of rat and human islets (30), whereas another group showed that ghrelin is present in pancreatic β-cells (35). There has also been a report that ghrelin does not co-express with any known islet hormone, although it is expressed in a quite prominent endocrine cell population in human fetal pancreas (34).

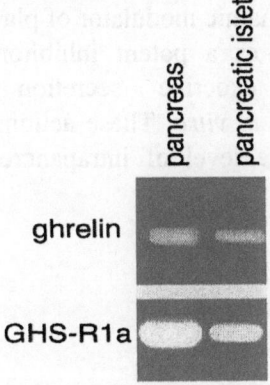

Figure 12-6. Electrophoretic analysis patterns of the RT-PCR products of ghrelin (top) and GHS-R1a (bottom) in rats.

GH Secretagogue receptor (GHS-R) type 1a (GHS-R1a) (the functional ghrelin receptor) mRNA is primarily expressed in the pituitary and, at a much lower level, in the thyroid gland, pancreas, spleen, myocardium, and adrenal gland (31). GHS-R1a mRNA was also detected in rat islets using reverse transcriptase-polymerase chain reaction (RT-PCR) (Fig. 12-6), but its localization within the islets remains unknown.

2.2   Functional role of ghrelin in the pancreas

Ghrelin and its receptor are expressed in rats islet cells, suggesting that ghrelin acts in some manner through GHS-R 1a in the islets. At a low dose of 10^{-12} M, ghrelin was found to stimulate insulin secretion from isolated rat pancreatic islets (30,36) and increases $[Ca^{2+}]_i$ in rat islet β-cells in the presence of stimulatory (8.3 mM) but not basal (2.8 mM) glucose concentrations (Fig. 12-7). In rats *in vivo*, iv administered ghrelin also increases insulin secretion (37). Compared to these results obtained with low concentrations, ghrelin at a higher concentration of 10^{-8} M has a less pronounced effect on both insulin release from isolated rat islets and $[Ca^{2+}]_i$ increase in rat islet β-cells (30). Also, ghrelin at a dose of 10^{-8} M has been reported to reduce the insulin response to the secretagogues glucose, arginine, and carbachol, which act on β-cells via different mechanisms (38). Plasma ghrelin concentrations in humans increase nearly two-fold immediately before meals and fall to basal levels within 1 hour after eating, a pattern reciprocal to that of insulin (39). In rats, plasma ghrelin concentrations are correlated with the duration of food intake during the dark phase only, and decrease by 26% in the 20 minutes following the end of a food intake period (40). Thus, there appears *in vivo* to be a negative association between ghrelin and insulin secretion. These findings suggest that the effect of ghrelin on insulin secretion may differ according to nutritional status. An effect of insulin on ghrelin secretion has also been reported (41). Insulin infusion into human subjects decreased plasma ghrelin concentration, indicating that insulin may be a dynamic modulator of plasma ghrelin. Ghrelin has been known to function as a potent inhibitor of pancreatic cholecystokinin (CCK)-induced exocrine secretion in anesthetized rats *in vivo* and in pancreatic lobes *in vitro*. These actions of ghrelin are indirect and may be exerted at the level of intrapancreatic neurons (42).

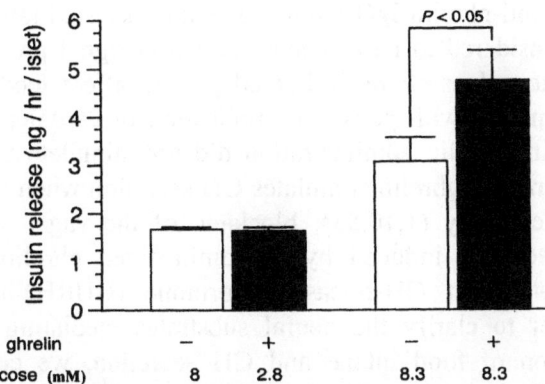

Figure 12-7. Effect of ghrelin on insulin release from rat islets. Ghrelin at 10^{-12} M stimulated insulin release in the presence of 8.3 mM glucose but not 2.8 mM glucose. The results are expressed as means ± standard error for 10 experiments in each group.

3. THE ROLE OF THE VAGUS NERVE IN GHRELIN'S ACTION

3.1 Functional role of the vagus nerve in ghrelin-induced feeding and GH secretion

The vagus nerve is one of the most complex nerves in the body, innervating nearly all of the thoracic and abdominal viscera. The afferent fibers of the vagus nerve form the major neuroanatomical linkage between the alimentary tract and the nucleus of the solitary tract in the hindbrain, where afferent input is integrated with descending hypothalamic input and ascending output to the hypothalamus is generated (43-46). Some of the information transmitted by the vagal afferents has been thought to control food intake. For example, CCK, a hormone released by gastrointestinal endocrine cells (47), suppresses food intake by acting on vagal afferent fibers (48,49).

As mentioned earlier, ghrelin stimulates food intake when administered peripherally or centrally (9,11,22-24,50,51). Central ghrelin probably affects feeding through the neural circuitry (22). On the other hand, work in our laboratory showed that ghrelin administered iv to free-feeding rats above a minimally active dose of 1.5 nmol increased food intake in a dose-dependent manner (Fig. 12-8). Ghrelin mRNA expression in the stomach increases after fasting, and returns to baseline levels after refeeding (14). Ghrelin concentration in the gastric vein also increases after fasting. In addition, iv administration of anti-ghrelin IgG reduces feeding in fasted rats. Therefore, ghrelin can be considered to be a potent starvation signal produced in the stomach. How, then, does stomach-derived ghrelin affect feeding? In rats treated by vagotomy or with perivagal application of capsaicin, a specific afferent neurotoxin, ghrelin administration did not stimulate feeding (Fig. 12-9) (52). Furthermore, ghrelin stimulates GH secretion when administered peripherally and centrally (1,10,53); blockade of the vagal afferent also attenuated GH secretion induced by iv administered ghrelin (52). This treatment did not affect GH-releasing hormone (GHRH)-induced GH secretion. In order to clarify the neural substrates mediating iv ghrelin-induced stimulation of food intake and GH secretion, we conducted an immunohistochemical analysis of Fos protein expression. Fos protein is the product of the protooncogene c-fos. This gene is believed to be a transcriptional modulator that is expressed in some neurons following their stimulation. The Fos protein, after being translated in the cytoplasm, binds together with another protein, Jun, in the nucleus, where it modulates the transcription of mRNAs encoding other protein products of the cell. Iv administered ghrelin induced Fos expression only in the arcuate nucleus (52,54). A double staining study showed that iv administered ghrelin induced Fos expression in 43% ± 5% of NPY-containing neurons and 15% ± 5% of GHRH-containing neurons. In addition, ghrelin administered iv to vagotomized or capsaicin-treated rats did not induce Fos expression in the arcuate nucleus. These data suggest that stomach-derived ghrelin acts predominantly to stimulate feeding and GH secretion via vagal afferent fibers.

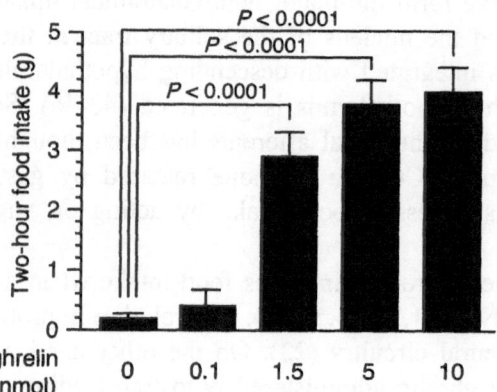

Figure 12-8. Two-hour food intake (mean ± standard error) of free feeding rats after a single iv administration of ghrelin (0.1-10 nmol).

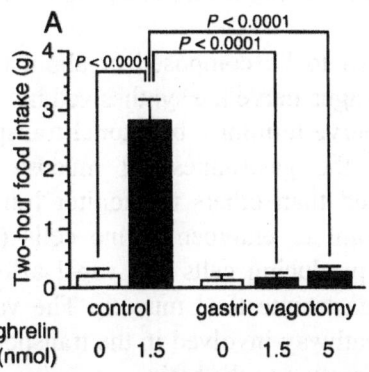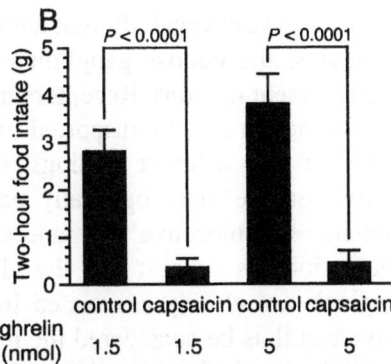

Figure 12-9. A: food intake of rats with gastric branch vagotomy after a single iv administration of ghrelin (1.5 or 5 nmol). Control rats underwent a sham operation; B: food intake of rats with perivagal capsaicin application after a single iv administration of ghrelin (1.5 or 5 nmol).

3.2 Ghrelin receptor expression in the vagus nerve

The notion that vagal afferent fibers are adapted to mediate iv ghrelin-induced stimulation of food intake is supported by two lines of technically independent investigation. First, GHS-R1a is expressed in vagal afferent neurons within the nodose ganglion located near the jugular foramen and then transported to the vagal afferent terminals (52). Second, electrophysiological data indicate that iv administered ghrelin significantly decreases gastric vagal afferent activity (52).

We investigated the expression of GHS-R1a in the vagal nodose ganglion using RT-PCR, direct sequencing, and *in situ* hybridization histochemistry. A GHS-R1a transcript product was found in an mRNA sample isolated from the vagal nodose ganglion. The DNA sequence of this PCR product was consistent with that of GHS-R1a. Signals specific for ^{33}P-labeled antisense probes for GHS-R1a were found in the afferent neuron-containing area of the nodose ganglion. The signals disappeared with the addition of an excess of unlabeled probe. Approximately 40% of neuronal cell bodies in the vagal nodose ganglion were positive for GHS-R1a probes.

Iv administered ghrelin (1.5 nmol) significantly decreases gastric vagal afferent activity. On the other hand, iv administration of des-acyl ghrelin, which lacks the *n*-octanoylation at serine 3 that is essential for ghrelin's receptor binding (1), does not induce food intake, stimulate GH secretion, or affect afferent activity. These findings imply the specificity of ghrelin in eliciting the observed electrophysiological effects on the gastric vagal afferent. In contrast, iv administration of CCK, which transmits a satiety signal to the nucleus of the solitary tract via vagal afferents, significantly increases gastric vagal afferent activity.

In rats, the nodose ganglion is known to be composed of about 6000 vagal afferent neurons. Receptors in the vagus nerve are synthesized in vagal afferent neurons and transported to the nerve terminals by axonal transport. Some of the afferent endings within the gastrointestinal mucosa and submucosa are more optimally positioned than others to monitor luminal contents and bioactive substances relevant to enteroendocrine cells (55). These findings suggest that the ghrelin-producing cells and vagal afferent terminals are closely associated in the gastrointestinal mucosa. The vagus nerve can thus be considered the major pathway involved in the transport of ghrelin's signals for starvation and GH secretion to the brain.

REFERENCES

1. Kojima M, Hosoda H, Date Y, Nakazato M, Matsuo H, Kangawa K. Ghrelin is a novel growth hormone releasing acylated peptide from stomach. Nature. 1999; 402:656-60.
2. Rindi G, Necchi V, Savio A, et al. Characterisation of gastric ghrelin cells in man and other mammals: studies in adult and fetal tissues. Histochem Cell Biol. 2002; 117:511-9.
3. Hayashida T, Nakahara K, Mondal MS, et al. Ghrelin in neonatal rats: distribution in stomach and its possible role. J Endocrinol. 2002; 173:239-45.
4. Dornonville de la Cour C, Bjorkqvist M, Sandvik AK, et al. A-like cells in the rat stomach contain ghrelin and do not operate under gastrin control. Regul Pept. 2001; 99:141-50.
5. van der Toorn FM, Janssen JA, de Herder WW, Broglio F, Ghigo E, van der Lely AJ. Central ghrelin production does not substantially contribute to systemic ghrelin concentrations: a study in two subjects with active acromegaly. Eur J Endocrinol. 2002; 147:195-9.
6. Date Y, Kojima M, Hosoda H, et al. Ghrelin, a novel growth-hormone-releasing acylated peptide, is synthesized in a distinct endocrine cell type in the gastrointestinal tracts of rats and humans. Endocrinology. 2000; 141:4255-61.
7. Sachs G, Zeng N, Prinz C. Physiology of isolated gastric endocrine cells. Annu Rev Physiol. 1997; 59:243-56.
8. Capella C, Bordi C, Monga G, et al. Multiple endocrine cell types in thyroid medullary carcinoma. Evidence for calcitonin, somatostatin, ACTH, 5HT and small granule cells. Virchows Arch A Pathol Anat Histol. 1978; 377:111-28.
9. Tschoep M, Smiley DL, Heiman ML. Ghrelin induces adiposity in rodents. Nature. 2000; 407:908-13.
10. Takaya K, Ariyasu H, Kanamoto N, et al. Ghrelin strongly stimulates growth hormone release in humans. J Clin Endocrinol Metab. 2000; 85:4908-11.
11. Wren AM, Seal LJ, Cohen MA, et al. Ghrelin enhances appetite and increases food intake in humans. J Clin Endocrinol Metab. 2001; 86:5992-5.

12. Tschoep M, Weyer C, Tataranni PA, Devanarayan V, Ravussin E, Heiman ML. Circulating ghrelin levels are decreased in human obesity. Diabetes. 2001; 50:707-9.

13. Nagaya N, Uematsu M, Kojima M, et al. Elevated circulating level of ghrelin in cachexia associated with chronic heart failure: relationships between ghrelin and anabolic/ catabolic factors. Circulation. 2001; 104:2034-8.

14. Toshinai K, Mondal MS, Nakazato M, et al. Upregulation of ghrelin expression in the stomach upon fasting, insulin-induced hypoglycemia, and leptin administration. Biochem Biophys Res Commun. 2001; 281:1220-5.

15. Shiiya T, Nakazato M, Mizuta M, et al. Plasma ghrelin levels in lean and obese humans and the effect of glucose on ghrelin secretion. J Clin Endocrinol Metab. 2002; 87:240-4.

16. Inui A. Ghrelin: an orexigenic and somatotrophic signal from the stomach. Nat Rev Neurosci. 2001; 8:551-60.

17. Tache Y, Goto Y, Gunion MW, Vale W, River J, Brown M. Inhibition of gastric acid secretion in rats by intracerebral injection of corticotropin-releasing factor. Science. 1983; 222:935-7.

18. Tache Y, Vale W, Brown M. Thyrotropin-releasing hormone--CNS action to stimulate gastric acid secretion. Nature. 1980; 287:149-51.

19. Tache Y, Yang H. Brain regulation of gastric acid secretion by peptides. Sites and mechanisms of action. Ann N Y Acad Sci. 1990; 597:128-45.

20. Okumura T, Yamada H, Motomura W, Kohgo Y. Cocaine-amphetamine-regulated transcript (CART) acts in the central nervous system to inhibit gastric acid secretion via brain corticotropin-releasing factor system. Endocrinology. 2000; 141:2854-60.

21. Masuda Y, Tanaka T, Inomata N, et al. Ghrelin stimulates gastric acid secretion and motility in rats. Biochem Biophys Res Commun. 2000; 276:905-8.

22. Nakazato M, Murakami N, Date Y, et al. A role for ghrelin in the central regulation of feeding. Nature. 2001; 409:194-8.

23. Shintani M, Ogawa Y, Ebihara K, et al. Ghrelin, an endogenous growth hormone secretagogue, is a novel orexigenic peptide that antagonizes leptin action through the activation of hypothalamic neuropeptide Y/Y1 receptor pathway. Diabetes. 2001; 50:227-32.

24. Wren AM, Small CJ, Ward HL, et al. The novel hypothalamic peptide ghrelin stimulates food intake and growth hormone secretion. Endocrinology. 2000; 141:4325-8.

25. Date Y, Nakazato M, Murakami N, Kojima M, Kangawa K, Matsukura S. Ghrelin acts in the central nervous system to stimulate gastric acid secretion. Biochem Biophys Res Commun. 2001; 280:904-7.

26. Sibilia V, Pagani F, Guidobono F, et al. Evidence for a central inhibitory role of growth hormone secretagogues and ghrelin on gastric acid secretion in conscious rats. Neuroendocrinology. 2002; 75:92-7.

27. Korbonits M, Bustin SA, Kojima M,et al. The expression of the growth hormone secretagogue receptor ligand ghrelin in normal and abnormal human pituitary and other neuroendocrine tumors. J Clin Endocrinol Metab. 2001; 86:881-7.

28. Lu S, Guan JL, Wand QP,et al: Immunocytochemical observation of ghrelin-containing neurons in the rat arcuate nucleus. Neurosci Lett. 2002; 321:157-60.

29. Mori K, Yoshimoto A, Takaya K, Hosoda K, et al: Kidney produces a novel acylated peptide, ghrelin. FEBS Lett. 2000; 486:213-6.

30. Date Y, Nakazato M, Hashiguchi S, et al. Ghrelin is present in pancreatic α-cells of humans and rats and stimulates insulin secretion. Diabetes. 2002; 51:124-129.

31. Gnanapavan S, Kola B, Bustin SA, et al: The tissue distribution of the mRNA of ghrelin and subtypes of its receptor, GHS-R, in humans. J Clin Endocrinol Metab. 2002; 87:2988-91.

32. Papotti M, Cassoni P, Volante M, Deghenghi R, Muccioli G, Ghigo E. Ghrelin-producing endocrine tumors of the stomach and intestine. J Clin Endocrinol Metab. 2001; 86:5052-9.

33. Creutzfeldt W. The incretin concept today. Diabetologia. 1979; 16:75-85.

34. Wierup N, Svensson H, Mulder H, Sundler F. The ghrelin cell: a novel developmentally regulated islet cell in the human pancreas. Regul Pept. 2002; 107:63-9.

35. Volante M, Allia E, Gugliotta P, et al: Expression of ghrelin and of the GH secretagogue receptor by pancreatic islet cells and related endocrine tumors. J Clin Endocrinol Metab. 2002; 87:1300-8.

36. Adeghate E, Ponery AS. Ghrelin stimulates insulin secretion from the pancreas of normal and diabetic rats. J Neuroendocrinol. 2002; 14:555-60.

37. Lee HM, Wang G, Englander EW, Kojima M, Greeley GH Jr. Ghrelin, a new gastrointestinal endocrine peptide that stimulates insulin secretion: enteric distribution, ontogeny, influence of endocrine, and dietary manipulations. Endocrinology. 2002; 143:185-90.

38. Egido EM, Rodriguez-Gallardo J, Silvestre RA, Marco J. Inhibitory effect of ghrelin on insulin and pancreatic somatostatin secretion. Eur J Endocrinol. 2002; 146:241-4.

39. Cummings DE, Purnell JQ, Frayo RS, Schmidova K, Wisse BE, Weigle DS. A preprandial rise in plasma ghrelin levels suggests a role in meal initiation in humans. Diabetes. 2001; 50:1714-9.

40. Tolle V, Bassant MH, Zizzari P,et al. Ultradian rhythmicity of ghrelin secretion in relation with GH, feeding behavior, and sleep-wake patterns in rats. Endocrinology. 2002; 143:1353-61.

41. Saad MF, Bernaba B, Hwu CM, et al. Insulin regulates plasma ghrelin concentration. J Clin Endocrinol Metab. 2002; 87:3997-4000.

42. Zhang W, Chen M, Chen X, Segura BJ, Mulholland MW. Inhibition of pancreatic protein secretion by ghrelin in the rat. J Physiol. 2001; 537:231-6.

43. Sawchenko PE. Central connections of the sensory and motor nuclei of the vagus nerve. J Auton Nerv Syst. 1983; 9:13-26.

44. van der Kooy D, Koda LY, McGinty JF, Gerfen CR, Bloom FE. The organization of projections from the cortex, amygdala, and hypothalamus to the nucleus of the solitary tract in rat. J Comp Neurol. 1984; 224:1-24.

45. Ritter S, Dinh TT, Friedman MI. Induction of Fos-like immunoreactivity (Fos-li) and stimulation of feeding by 2,5-anhydro-D-mannitol (2,5-AM) require the vagus nerve. Brain Res. 1994; 646:53-64.

46. Schwartz MW, Woods SC, Porte D Jr, Seeley RJ, Baskin DG. Central nervous system control of food intake. Nature. 2000; 404:661-71.

47. Ivy AC, Oldenberg E. A hormone mechanism for gallbladder contraction and evacuation. Am J Physiol. 1928; 86:599-613.

48. Ritter RC, Ladenheim EE. Capsaicin pretreatment attenuates suppression of food intake by cholecystokinin. Am J Physiol. 1985; 248:R501-4.

49. South EH, Ritter RC. Capsaicin application to central or peripheral vagal fibers attenuates CCK satiety. Peptides. 1988; 9:601-12.

50. Kamegai J, Tamura H, Shimizu T, Ishii S, Sugihara H, Wakabayashi I. Central effect of ghrelin, an endogenous growth hormone secretagogue, on hypothalamic peptide gene expression. Endocrinology. 2000; 141:4797-800.
51. Asakawa A, Inui A, Kaga T, et al. Ghrelin is an appetite-stimulatory signal from stomach with structural resemblance to motilin. Gastroenterology. 2001; 120:337-45.
52. Date Y, Murakami N, Toshinai K, et al. The role of the gastric afferent vagal nerve in ghrelin-induced feeding and growth hormone secretion in rats. Gastroenterology. 2002; 123:1120-8.
53. Date Y, Nakazato M, Murakami N, Kojima M, Kangawa K, Matsukura S. Ghrelin acts in the central nervous system to stimulate gastric acid secretion. Biochem Biophys Res Commun. 2001; 280:904-7.
54. Hewson AK, Dickson SL. Systemic administration of ghrelin induces Fos and Egr-1 proteins in the hypothalamic arcuate nucleus of fasted and fed rats. J Neuroendocrinol. 2000; 12:1047-9.
55. Grundy D, Acretcherd T. Sensory afferents from the gastrointestinal tract. In: Schultz SG, ed. Handbook of pyhsiology; the gastrointestinal system. Vol I. Motility and circulation. New York: Oxford University Press. 1989; 593-620.

Chapter 13

A NEW INTERACTION TO COME: GHRELIN AND STEROID HORMONES

Uberto Pagotto, Alessandra Gambineri, Valentina Vicennati & Renato Pasquali
Endocrinology Unit, Department of Internal Medicine and Gastroenterology, Center of Applied Biomedical Research, S.Orsola-Malpighi Hospital, Via Massarenti, 9 40138 Bologna, Italy

Abstract: Apart from the well known effect of ghrelin in the stimulation of pituitary hormones and in the control of food intake and energy balance, ghrelin has been hypothesized to play an important role in many other endocrine and nonendocrine functions. In this context, gonadal and adrenal glands have very recently been exposed as potential targets of ghrelin action, based on a high number of binding sites found in both female and male gonads and in adrenals. In addition, the intratissutal presence of this hormone shown in the glands mentioned above of different species make it possible to hypothesize an autocrine/paracrine ghrelin role. This hormone is able to modulate *in vitro* steroidogenesis in gonads and to stimulate proliferation in adrenal cortex. On the other hand, steroids *per se* are able to modify *in vivo* ghrelin concentrations. The changes of basal circulating ghrelin levels related to the pathological variations of sexual steroid hormones as observed in patients affected by polycystic ovary syndrome and hypogonadism make it possible to include sexual hormones as one of the main important regulators of ghrelin synthesis or secretion. This new evidence provides emphasis to the concept that ghrelin should also be considered as a new powerful link between reproduction and the metabolic processes.

Key words: adrenal, ovary, testis, steroids, gender

1.   INTRODUCTION

The connection between ghrelin and steroid hormone appears to be based on a reciprocal interaction. On one hand ghrelin is able to directly impact the most relevant functions of gonads and adrenals, by activating its own receptors. On the other hand evidence based on a *in vivo* studies clearly points to a strong ability of steroids, in particular the sexual hormones, to considerably modulate ghrelin circulating levels. The aim of this chapter will be to summarize the findings which make it possible to support the existence of this reciprocal interaction.

2.   GHRELIN EXPRESSION

2.1   Testis

According to the mRNA quantification provided by the study of the group of Korbonits, human ghrelin transcripts are abundantly expressed in human testis (1). The same holds true in rats as demonstrated by the findings of Dieguez group's (2,3). Ghrelin expression displays the highest value in adult rat testis, however it is also detectable throughout postnatal development (3). Ghrelin is mainly located in mature Leydig cells. Interestingly, this expression is under the regulation of luteinizing hormone (LH) as demonstrated by experiments in which administration of chorionic gonadotropin (CG - an agonist of LH) restored the absence of ghrelin expression induced by hypophysectomy, whereas follicle stimulating hormone (FSH) was not able to induce it (3). Notably, a ghrelin gene-derived transcript has been isolated in mouse testis. The function of this splice variant is at present unknown, but interestingly, the expression profile during development mimics that of full-length ghrelin (4).

2.2   Ovary

Dieguez and colleagues have provided definitive evidence that the ovary may be considered an important site of production of ghrelin (5). In rats, ghrelin has been found in the luteal compartment of the ovary (5). Importantly, ghrelin expression peaks in connection with the functional phase of the corpus luteum, whereas during corpus luteum formation and regression a reduced expression has been detected. These findings allowed Dieguez to hypothesize that ghrelin may have a potential functional role in the regulation of luteal development (5). Studies of the human ovary by the same group produced similar findings (6). In fact, ghrelin immunoreactivity has been found in luteal cells in both a young and mature phase, when the corpus luteus is at the maximal activity for progesterone secretion. On the other hand, intense immunopositivity for ghrelin has also been found in hilus interstitial cells (6). This cell type maybe considered the female counterpart of the testicular Leydig cells. In fact, both cell types are characterized by the presence of the crystals of Reinke, and are under the control of LH for the synthesis and secretion of androgens.

2.3 Adrenal

Ghrelin mRNA and protein have been detected in the rat adrenal cortex, and the amount of ghrelin immunoreactivity measured by radioimmunoassay was found to be higher than that detected in testis (7). On the other hand, another study has failed to show any ghrelin transcript in the rat adrenal gland (8). The reason for this discrepancy has been attributed to the different strain of rats examined. Conversely, in humans, ghrelin mRNA has been shown to be less expressed than in testes (1). However, a very recent study has found sizeable expression of ghrelin mRNA and protein in a huge series of human adrenal cortex tissues pointing to an involvement of ghrelin in the autocrine/paracrine regulation of the adrenal gland functions (9).

3. GHRELIN RECEPTORS

Many studies have already characterized the structure and the binding properties of ghrelin receptors (10). An alternative splicing gives rise to the two different variants of ghrelin receptor: the full-length type 1a and the truncated form named type 1b lacking the transmembrane domains 6 and 7. Apparently, only type 1a is functionally active, whereas to date type 1b seems to be devoid of signal transduction activity (10).

3.1 Testis

Studies at mRNA levels have showed the occurrence in rat testes of a progressive shift during pubertal development toward the preferential expression of type 1a receptor (11). In the same animal model type 1a receptor was localized within adult testes in Sertoli cells and interstitial Leydig cells. This expression detected at transcript level was confirmed at protein level by using an antibody raised against the C-terminal fragment of type 1a receptor. Importantly, the mRNA expression of both receptors seems to be under the close control of ghrelin itself and FSH. Both variants have been detected in rat seminiferous epithelium with relevant changes in the expression pattern in a stage-specific manner. This finding suggests that the testes may be under dynamic control by ghrelin. A further confirmation of the testes as a key target organ for ghrelin action has been given by a study in human tissues in which human testes have been represented as one of the organs displaying a higher number of binding sites as compared to a large number of endocrine and nonendocrine tissues (12). Testosterone may regulate ghrelin receptor expression in rat testes (11). This makes it possible to hypothesize a mutual interaction between ghrelin and steroids.

3.2   Ovary

A very extensive study has been performed in order to investigate the presence of ghrelin receptors in the human ovary. Immunohistochemical studies using an antibody specifically recognizing a C-terminal fragment of the human ghrelin receptor type 1a showed a wide expression pattern of this receptor with signals detected in oocytes as well as in follicular cells and in luteal cells during the entire morphological evolution. Notably, the receptor expression paralleled the follicle development, being very strong in granulosa and theca layers of antral follicles (6). This is suggestive of a potential relationship between follicle growth and ghrelin type 1a receptor (6).

3.3   Adrenal

In various species, the adrenal gland exhibits a moderate mRNA signal for ghrelin type 1a receptor that undergoes a homologous down-regulation. In the rat adrenal gland as well in the human adrenal gland, autoradiography studies demonstrated abundant [^{125}I]ghrelin binding sites in the outer portion of the cortex such as in the zona glomerulosa and in the outer zona fascicolata (7,9). Only low binding sites have been found in the adrenal medulla (7).

4.   PUTATIVE ACTION OF GHRELIN AT THE LEVEL OF STEROID-PRODUCING GLANDS

At present, all studies investigating the potential role of ghrelin in modulating physiological functions of gonads and adrenal glands have only been performed in animals.

4.1   Testis

Ghrelin has the ability to directly reduce, in a dose-dependent manner, the secretion of testosterone induced by the CG and the cAMP. This effect has been explained by the concomitant down-regulation of key enzymes involved in the steroidogenesis such as the steroid acute regulatory protein (StAR), the P450 cholesterol side-chain cleavage (P450scc), the 3β-hydroxy-steroid dehydrogenase (3β-HSD) and the 17β-hydroxy-steroid dehydrogenase type III enzymes (2).

4.2   Ovary

No studies have been conducted so far with the aim of investigating the direct role of ghrelin in regulating steroidogenic enzymes in the ovaries.

4.3   Adrenal

Two studies have reached similar results concluding that ghrelin seems to be unable to evoke any change in the basal secretion of the most important steroids produced by the rat adrenal gland inclusive of corticosterone and aldosterone. This lack of effect has been demonstrated both in basal condition and after a co-incubation with appropriate stimuli such as adrenocorticotropic hormone ACTH or angiotensin II (7,8). As expected, no changes have been observed after ghrelin stimulation in some upstream elements of steroidogenesis as the StAR, the P450scc and 3β-HSD. On the contrary, starting from the finding of abundant expression of ghrelin receptors in the cells composing the zona glomerulosa and bearing in mind that this is the layer involved in the maintenance of adrenal growth, Nussdorfer's group (7) has initiated a series of experiments in which the putative growth stimulatory ability of ghrelin has been tested. In fact, ghrelin was able to stimulate the growth of zona glomerulosa cells by an activation of ghrelin type 1a receptor. Notably, in a series of experiments using inhibitors of specific pathway steps, the proliferative effect has been attributed to an involvement of tyrosine-kinase/mitogen-activated protein kinase signaling pathway (7). One can conclude that ghrelin should be included in the long list of peptides that may affect the growth of the adrenal gland in an autocrine or paracrine role.

5.   GHRELIN AND STEROIDS IN HUMANS

5.1   Gender effect

The first publications comparing ghrelin circulating levels between females and males did not show any difference when gender was taken into account (13). However, in a recent study, mean 24 hours ghrelin concentrations were 3-fold higher in women in the follicular phase of the menstrual cycle when compared to men of the same age, making it possible to hypothesize that sexual steroids may importantly influence ghrelin plasma levels (14).

5.2   Ghrelin in pregnancy

Pregnancy is a physiological condition characterized by an important alteration in sexual steroid milieu, and therefore represents an important state in which ghrelin levels may be speculated to be influenced by changes in steroid levels. In fact, the maternal plasma concentration of ghrelin has been shown to be importantly reduced in normal pregnant women at the third trimester when compared to normal non-pregnant women (15). At present there are no studies in which the interconnection between steroids and ghrelin has been investigated. Nevertheless, it has been hypothesized that ghrelin may have an important role as mediator in cardiovascular control during pregnancy or possibly involved in the regulation of fetal-placental circulation (16).

5.3   Ghrelin in polycystic ovary syndrome

This disease, affecting at least 7.5% of women of reproductive age, is characterized by a hyperandrogenic state and chronic anovulation (17). Moreover, most of these patients present insulin resistance and other characteristics of the so-called "metabolic syndrome". Many groups have investigated the putative alteration of ghrelin circulating levels in patients affected by polycystic ovary syndrome (PCOS) in consideration of the fact that obesity is frequently associated with this syndrome (18), bearing in mind that ghrelin plasma levels are reduced in diseases characterized by concomitant obesity (10).

Three groups have found that PCOS patients have lower circulating ghrelin levels than healthy lean or simple obese patients (19-22). However, while a close negative correlation with insulin resistance was found by all three groups, it is important that our study and that performed by Moran and colleagues also found a close correlation with circulating androgens (20-22). In particular, we found that a marked negative correlation exists between ghrelin and andostenedione in simple obese women and in obese women affected by PCOS. Importantly, this correlation was also maintained in both populations after six months' therapy with metformin, an insulin-sensitizer drug, added to a low calorie diet (20). Even if insulin resistance significantly decreased during the administration of this therapy, neither ghrelin nor the main androgens significantly changed after therapy. This finding has been the first to hypothesize that the relationship between ghrelin and androgens may be more relevant than that between this hormone and insulin resistance, as proposed by most of the studies. A subsequent study by our group further substantiated this assumption. In fact, we studied a new group of 14 obese PCOS on a hypocaloric diet, randomly assigned to an antiandrogen treatment (flutamide) or to placebo for six months. At the end of the study, a significant reduction of androgen levels was only observed in the group assigned to flutamide treatment, while insulin resistance changed similarly and significantly in both groups. Importantly, plasma ghrelin levels significantly increased only following treatment with flutamide, whereas ghrelin remained unchanged in the placebo group (Figure 13-1A). Moreover, a negative correlation between changes in ghrelin and changes in androgen levels was observed in the active treated group (21). In the same study we also confirmed the presence of a positive correlation between changes in plasma ghrelin and those of insulin sensitivity indexes. This correlation, however, was insignificant after adjusting for androgen changes, thus suggesting that androgens may be primarily involved in the regulation of plasma ghrelin levels.

5.4　Ghrelin in male hypogonadism

Male hypogonadism of both central and peripheral origin is a pathological condition associated with low levels of circulating androgens. This status is frequently associated with overweight and insulin resistance. To further investigate the relationship between sex hormones and ghrelin we investigated a group of hypogonadal men compared to body mass index and body fat distribution matched eugonadal men and a group of normal weight individuals. In addition, in a second part of the study, the group of hypogonadal men was tested for ghrelin circulating levels before and after the substitution therapy with testosterone. In basal conditions, we found that circulating ghrelin levels in the hypogonadal group were significantly lower than those observed in both controls groups. Moreover, a highly significant positive correlation was also present between ghrelin and testosterone levels throughout all three groups involved in our study. Interestingly, after 6 months' therapy hypogonadal men restored ghrelin values to the normal range regardless of changes in body mass and insulin sensitivity (Figure 13-1B).

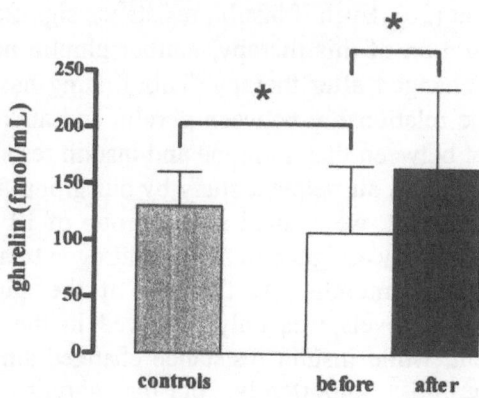

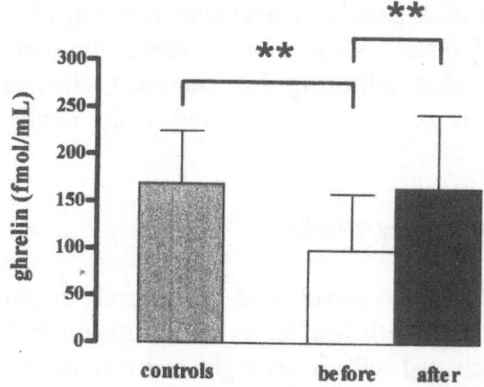

Figure 13-1. Ghrelin levels (mean±standard error of the mean) before and after 6 months therapy with flutamide or testosterone in women with polycystic ovary syndrome (A) and in hypogonadal men (B). Controls are age- and sex-matched. *: p<0.05; **: p<0.01.

These data further demonstrated that changes in circulating androgens and not variations in body composition or insulin sensitivity are the dominant factors in the regulation of ghrelin plasma levels (23).

6.   CONCLUSIVE REMARKS AND FUTURE PERSPECTIVES

The whole body of the data in humans seems to indicate that circulating steroids should be considered as important factors in the regulation of circulating ghrelin levels (Figure 13-2).

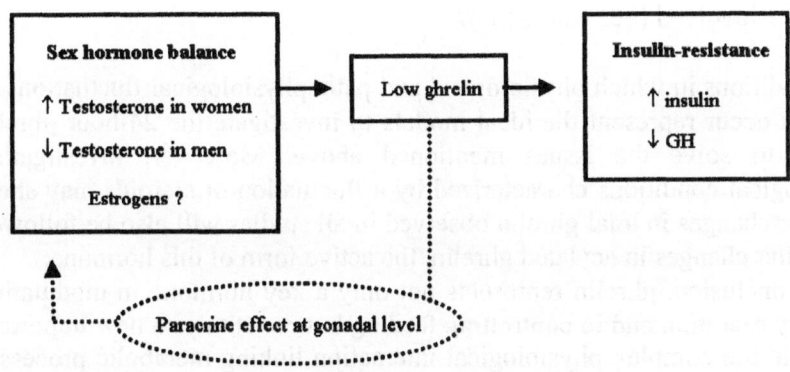

Figure 13-2. The sex hormones/ghrelin system in males and females.

These findings are abundantly substantiated by a series of *in vitro* reports in which gonads and adrenals have been shown to be not only the site of production but also the target of ghrelin action.

Although the *in vivo* data still do not allow definitive conclusions to be drawn, some speculations could be proposed. It seems that steroids may play a role more as fine-tune regulators of ghrelin secretion than as strong and rapid stimulators like those factors, still unknown, responsible for the dramatic fluctuations of ghrelin during eating. The normalization of the amount of plasma ghrelin after restoration of normal androgen status either in females or in males and the increase of circulating ghrelin levels associated with a particular menstrual phase seems to suggest that ghrelin may represent a new hormonal link between reproduction and metabolic processes.

There are still some open questions which should be solved in the near future:

1)   Does intratissutal ghrelin have a general endocrine effect contributing in this way to the total amount of circulating ghrelin

or, likely, does the intra-gonadal and intra-adrenal ghrelin display an eminent paracrine/autocrine role?

2)   Does ghrelin directly interact with steroids or is this interaction supported by any still unknown mediator?

3)   Does ghrelin only interact with peripheral target steroid-producing cells or does this relationship also involve the ability of this hormone to influence pituitary or hypothalamic hormonal secretion, regulating in turn the peripheral target glands, as has been hypothesized by some studies (24)?

4)   Do androgens directly regulate ghrelin levels or are other factors involved (i.e., estrogens)?

Conditions in which physiological and pathophysiological fluctuations of steroids occur represent the ideal models to investigate the 24-hour ghrelin profile to solve the issues mentioned above. Moreover, investigating pathological conditions characterized by a fluctuation of steroids may show whether changes in total ghrelin observed in all studies will also be followed by similar changes in acylated ghrelin, the active form of this hormone.

In conclusion, ghrelin represents not only a key hormone in modulating pituitary secretion and in controlling feeding but constitutes a new important player in the complex physiological interaction linking metabolic processes and reproductive and stress-response functions.

ACKNOWLEDGEMENTS

We thank Dr. Rosaria de Iasio for the sample measurements.

REFERENCES

1.   Gnanapavan S, Kola B, Bustin SA, et al. The tissue distribution of the mRNA of ghrelin and subtypes of its receptor, GHS-R, in humans. J Clin Endocrinol Metab. 2002; 87:2988-91.
2.   Tena-Sempere M, Barreiro ML, Gonzalez LC, et al. Novel expression and functional role of ghrelin in rat testis. Endocrinology. 2002; 143:717-25.
3.   Barreiro ML, Gaytán F, Caminos JE, et al. Cellular location and hormonal regulation of ghrelin expression in rat testis. Biol Reprod. 2002; 67:1768-76.
4.   Tanaka M, Hayashida Y, Nakao N, Nakai N, Nakashima K. Testis-specific and developmentally induced expression of a ghrelin gene-derived transcript that encodes a novel polypeptide in the mouse. Biochem Biophys Acta. 2001; 1522:62-5.
5.   Caminos JE, Tena-Sempere M, Gaytan F, et al. Expression of ghrelin in the cyclic and pregnant ovary. Endocrinology. 2003; 144:1594-602.

6. Gaytán F, Barreiro ML, Chopin LK, et al. Immunolocalization of ghrelin and its functional receptor, the type 1a GH-secretagogue receptor, in the cyclic human ovary. J Clin Endocrinol Metab. 2003; 88:879-87.
7. Andreis PG, Malendowicz LK, Trjter M. et al. Ghrelin and growth hormone secretagogue receptor are expressed in the rat adrenal cortex: evidence that ghrelin stimulates the growth, but not the secretory activity of adrenal cells. FEBS Lett. 2003; 536:173-9.
8. Barreiro ML, Pinilla L, Aguilar E, Tena-Sempere M. Expression and homologous regulation of GH secretagogue receptor mRNA in rat adrenal gland. Eur J Endocrinol. 2002; 147:677-88.
9. Tortorella C, Macchi C, Spinazzi R, Malendowicz LK, Trejter M, Nussdorfer GG. Ghrelin, an endogenous ligand for the growth hormone-secretagogue receptor, is expressed in the human adrenal cortex. Int J Mol Med. 2003; 12:213-7.
10. Muccioli G, Tschop M, Papotti M, Deghenghi R, Heiman M, Ghigo E. Neuroendocrine and peripheral activities of ghrelin: implications in metabolism and obesity. Eur J Pharmacol. 2002; 440:235-54.
11. Barreiro ML, Souminen JS, Gaytàn F, et al. Developmental, stage-specific and hormonally regulated expression of growth hormone secretagogue receptor messenger ribonucleic acid in rat testis. Biol Reprod. 2003; 68:1631-40.
12. Papotti M, Ghe C, Cassoni P, et al. Growth hormone secretagogue binding sites in peripheral human tissues. J Clin Endocrinol Metab. 2000; 85:3803-7.
13. Ariyasu H, Takaya K, Tagami T, et al. Stomach is a major source of circulating ghrelin, and feeding state determines plasma ghrelin-like immunoreactivity levels in humans. J Clin Endocrinol Metab. 2001; 86:4753-8.
14. Barkan AL, Dimaraki EV, Jessup SK, et al. Ghrelin secretion in humans is sexually dimorphic, suppressed by somatostatin, and not affected by the ambient growth hormone levels. J Clin Endocrinol Metab. 2003; 88:2180-4.
15. Makino Y, Hosoda H, Shibata K, et al. Alteration of plasma ghrelin levels associated with the blood pressure in pregnancy. Hypertension. 2002; 39:781-4.
16. Gualillo O, Caminos JE, Blanco M, et al. Ghrelin, a novel placental-derived hormone. Endocrinology. 2001; 142:788-94.
17. Franks S. Polycystic ovary syndrome. N Engl J Med. 1995; 333: 853-61.
18. Gambineri A, Pagotto U, Pelusi C, Vicennati V, Pasquali R. Obesity and the polycystic ovary syndrome. Int J Obes. 2002; 26: 883-96.
19. Schöfl C, Horn R, Schill T, Schlösser HW, Müller MJ, Brabant G. Circulating ghrelin levels in patients with polycystic ovary syndrome. J Clin Endocrinol Metab. 2002; 87:4607-10.
20. Pagotto U, Gambineri A, Vicennati V, Heiman ML, Tschöp M, Pasquali R. Plasma ghrelin, obesity, and the polycystic ovary syndrome: correlation with insulin resistance and androgen levels. J Clin Endocrinol Metab. 2002; 87: 5625-9.
21. Gambineri A, Pagotto U, Tschöp M, et al. Anti-androgen treatment increases circulating ghrelin levels in obese women with polycystic ovary sindrome. J Endocrinol Invest. 2003; 26: 493-8.
22. Moran L, Noakes M, Clifton P, Wittert G, Norman RJ. Ghrelin secretion following weight loss and different composition in PCOS. Evidence of impaired response. Program of the 85th Annual Meeting of the Endocrine Society, Philadelphia, PA, 2003; p 84 (abstract).
23. Pagotto U, Gambineri A, Pelusi C, et al. Testosterone replacement therapy restores normal ghrelin in hypogonadal men. J Clin Endocrinol Metab. 2003; 88:4139-43.

24. Furuta M, Funabashi T, Kimura F. Intracerebroventricular administration of ghrelin rapidly suppresses pulsatile luteinizing hormone secretion in ovariectomized rats. Biochem Biophys Res Comm. 2001; 288:780-5.

Chapter 14

CIRCULATING GHRELIN LEVELS IN PATHOPHYSIOLOGICAL CONDITIONS

David E. Cummings & Joost Overduin
Department of Medicine, Division of Metabolism, Endocrinology and Nutrition, University of Washington, VA Puget Sound Health Care System, Seattle, WA, USA

Abstract: Circulating levels of ghrelin, are regulated by short-term factors pertaining to food ingestion and longer-term factors pertaining to body weight. The short-term, or prandial, regulation is manifest as marked increases in ghrelin levels before each meal and decreases after food is consumed. This temporal pattern, implicates ghrelin as a contributor to pre-meal hunger and the initiation of individual meals. Long-term, body weight-related regulation of ghrelin results in a negative relationship between ghrelin levels and numerous measures of body size. This association and several other findings suggest that ghrelin may participate not only in the short-term control of meal patterns but also in overall body-weight regulation. Ghrelin fulfills all of the established criteria to be an "adiposity signal" that senses the status of body-fat stores and communicates this information to the brain, which activates compensatory changes in appetite and energy expenditure designed to maintain homeostasis. Ghrelin levels vary in response to alterations in energy balance, increasing with weight loss and thus potentially contributing to the compensatory hyperphagia triggered by negative energy balance. The opposite is true for weight gain. Most of the alterations of circulating ghrelin levels in pathophysiological conditions can be viewed as adaptive responses to the body-weight perturbations in these disorders. The only known exception is Prader-Willi syndrome, a condition in which hyperghrelinemia may play a primary, causal role driving hyperphagia and obesity. These patients represent logical subjects in whom first to test the efficacy of ghrelin-blocking agents to treat obesity. Low fat diets promote modest weight loss without triggering the normal compensatory increase in ghrelin levels, and gastric bypass surgery can suppress (or at least constrain) ghrelin levels in most cases. The impact of these interventions on circulating ghrelin may contribute to their weight-reducing effects.

Key words: obesity, body-weight regulation, Prader-Willi syndrome, gastric bypass, anorexia

1.   INTRODUCTION

Circulating levels of the orexigenic hormone, ghrelin, are regulated by short-term factors pertaining to food ingestion and longer term factors pertaining to body weight. The short-term, or prandial, regulation is manifest as marked increases in ghrelin levels before each meal and decreases after food is consumed (1-4). This temporal pattern, together with several other observations (reviewed in Ref. 5), implicates ghrelin as a possible contributor to pre-meal hunger and the initiation of individual meals. Long-term, body weight-related regulation of ghrelin results in a negative relationship between ghrelin levels and numerous measures of body size (6-12). This association and several other findings suggest that ghrelin may participate not only in the short-term control of meal patterns but also in overall body weight regulation (5). Ghrelin fulfills all of the established criteria to be an "adiposity signal" that senses the status of body fat stores and communicates this information to the brain, which responds to alterations in energy stores with compensatory changes in appetite and energy expenditure designed to resist such changes (13). As predicted for an adiposity signal, ghrelin levels vary in response to alterations in energy balance, increasing with weight loss and thus potentially contributing to the compensatory hyperphagia triggered by negative energy balance (5). The opposite is true for weight gain (14).

Most of the alterations of circulating ghrelin levels that accompany pathophysiological conditions are consistent with ghrelin's proposed role as an adiposity signal, and can be viewed as adaptive responses to the body weight perturbations that are caused by these disorders. The only known exception is Prader-Willi syndrome (PWS), a condition in which hyperghrelinemia may play a primary, causal role driving hyperphagia and obesity. People with this disorder represent logical subjects in whom first to test the efficacy of ghrelin-blocking agents to treat obesity. Low fat diets promote modest weight loss without triggering the normal compensatory increase in ghrelin levels, and gastric bypass surgery can suppress (or at least constrain) ghrelin levels in most cases. The impact of these interventions on circulating ghrelin may contribute to their weight reducing effects.

2.   PRANDIAL GHRELIN REGULATION AND POSSIBLE ROLES FOR GHRELIN IN MEAL INITIATION

Meal time hunger is a common, daily experience, yet the molecular determinants of this sensation remain enigmatic despite decades of research to identify them. The following observations from rodent studies are consistent with the hypothesis that ghrelin, the only known circulating orexigen, contributes to preprandial hunger and participates in meal initiation:

a) the majority (probably about 2/3) of circulating ghrelin is produced by the stomach, an organ sensitive to short-term fluxes in nutrient balance (8,15-18);

b) as predicted for a meal initiator, ghrelin levels increase with fasting and are suppressed within minutes by re-feeding or enteral infusions of nutrients but not water (1,10,19,20);

c) exogenous ghrelin triggers eating in rodents when administered at times of minimal spontaneous food intake (19-22);

d) ghrelin's orexigenic actions are extremely rapid and short-lived, as required for a signal influencing individual meal-related behavior (19-22);

e) detailed analysis of meal patterns following ghrelin injections reveals that the primary orexigenic effect of ghrelin is to decrease the latency to feed, leading to one extra bout of feeding that occurs immediately after ghrelin administration (23);

f) ghrelin stimulates gastric motility and acid secretion, both of which increase in anticipation of meals (20,24);

g) the most clearly documented targets of ghrelin action in the brain – hypothalamic neuropeptide Y (NPY)/agouti-related protein (AgRP) neurons (5) – are implicated in the central control of meal initiation, because levels of these neuropeptides increase at times of maximal feeding in rodents (25). In contrast, the expression of other neuropeptides involved in energy balance is relatively constant throughout the day.

The hypothesis that circulating ghrelin is a physiologic meal initiator predicts that levels should rise before, and fall after, every meal, and peak concentrations should be sufficiently high to stimulate appetite. Consistent with this model, human ghrelin levels are indeed rapidly suppressed by nutrient ingestion (2,10,19), and 24-hour plasma profiles reveal marked preprandial increases and postprandial decreases associated with every meal (1). Ghrelin levels increase before meals to values that have been shown to stimulate appetite and food intake when generated by peripheral ghrelin administration in humans and rodents (26,27), suggesting that these levels are sufficient to play a physiologic role in normal pre-meal hunger. In people and animals habituated to various fixed meal patterns, preprandial ghrelin surges are observed before every meal, occurring as many times per day as meals occur (1,3,4,28). Finally, the depth and duration of prandial ghrelin suppression is related dose-dependently to the caloric load of the meal (29). In other words, large meals suppress ghrelin more thoroughly than small meals, just as they do with hunger. Together, these observations present a compelling picture of ghrelin as a meal initiator. The data, however, are circumstantial, and definitive loss-of-function experiments with ghrelin-blocking agents or genetic ablations are required to prove or disprove this hypothesis.

The mechanisms by which nutrient ingestion suppresses ghrelin levels are unknown. We have found in rats that ghrelin cells in the stomach and duodenum – the principal sites of production – are not affected directly by nutrients in the gastrointestinal lumen (Ref. 30; Overduin J, Frayo RS and Cummings DE, unpublished results). These observations suggest either that a nutrient-sensing mechanism located farther downstream in the intestine regulates ghrelin secretion indirectly, or that this is accomplished via post-absorptive events, and we are actively investigating these possibilities.

3.   BODY WEIGHT-RELATED GHRELIN REGULATION AND MODEL OF GHRELIN AS AN ADIPOSITY SIGNAL

In addition to being influenced acutely by ingested nutrients, ghrelin levels appear also to be regulated chronically by components of body weight, the exact nature of which have yet to be defined. Ghrelin levels correlate (inversely) with numerous measures of body-energy stores (6-12). This property of ghrelin is one of the characteristics of "adiposity signals" that participate in long-term body weight regulation (31, see also Chapter 7). Such peripheral signals communicate the status of body fat stores to the brain, and fluctuations in them trigger compensatory changes in food intake and energy expenditure that resist alterations in body weight (13). An adiposity signal that participates in energy homeostasis should manifest the following qualities:

a) it should circulate in proportion to body energy stores, and consequently, should be modulated reciprocally by increases or decreases in these stores;

b) it should gain access to the brain and interact there with receptors and signal transduction systems in neurons that participate in body weight regulation;

c) exogenous administration of the compound should alter food intake and/or energy expenditure, and chronic infusions should change body weight (or at least fat mass);

d) blockade of the signal should render the opposite effects.

Until recently, leptin and insulin were the only hormones shown to satisfy these criteria (13).

Somewhat surprisingly for a gut hormone, ghrelin fulfills all of the established criteria for peripheral adiposity signals, and may thus be a unique orexigenic counterpart to leptin and insulin in long-term body weight regulation. Circulating ghrelin levels correlate inversely with the size of energy stores over a very wide weight range, extending from emaciated victims of anorexia nervosa to super-obese humans and rodents with genetically absent leptin signaling (6-12). Although humoral ghrelin may not efficiently cross the blood-brain barrier (32), central targets of ghrelin action are located in regions of the hypothalamus and brainstem that are not well protected by this barrier and are known to regulate energy homeostasis (5, 23). The best documented of these targets are neurons in the arcuate nucleus of the hypothalamus that coexpress NPY and AgRP, both prototypic anabolic neuropeptides that promote positive energy balance (31). Almost all arcuate NPY/AgRP neurons express the ghrelin receptor (33), and ghrelin clearly activates these cells, as demonstrated by increases their firing rate and their expression of *c-fos* (a marker of neuronal activity), NPY, and AgRP (20, 21, 34-37). Furthermore, pharmacological blockade of either NPY or AgRP signaling attenuates the orexigenic actions of ghrelin (21,38). Peripheral or central ghrelin administration stimulates short-term food intake as potently as does any known agent (20,26), and chronic or repeated infusions increase body weight (19-21,27). In addition to stimulating food intake, ghrelin can decrease energy expenditure (20), fat catabolism (19), adipocyte apoptosis, sympathetic nervous system activity (39), body temperature (40), and locomotor activity (Tang-Christensen M and Tschöp M, unpublished results). In other words, ghrelin affects all aspects of the energy homeostasis system in a concerted manner to promote weight gain. The most compelling evidence of a critical role for ghrelin in body weight regulation would be a demonstration that blocking the endogenous hormone causes weight loss. Although this pivotal issue is not yet definitively addressed, all five studies published to date on the topic report that disruption of ghrelin signaling (using genetic techniques, anti-ghrelin antibodies, or ghrelin-receptor antagonists) decreases spontaneous food intake, leading to weight loss in the longer studies (21,41-44). Finally, a mutation in the human preproghrelin gene is reported to be associated with protection against obesity and related metabolic sequelae (45).

If circulating ghrelin acts as a barometer of bodily fuel stores and participates in energy homeostasis, its levels should change reciprocally in response to increases or decreases in body weight. Specifically, levels of this orexigen should rise with weight loss and fall with weight gain, consistent with adaptive responses to these perturbations (13). Indeed, such alterations in ghrelin levels have been reported in the context of numerous and varied conditions of body weight change. Ghrelin levels increase in response to weight loss resulting from low calorie diets (46,47), mixed lifestyle modifications (48), cancer anorexia (47,49), anorexia nervosa (8, 10,11), and chronic failure of the heart (50), liver (51), or kidneys (52). Conversely, a trend toward decreased ghrelin levels was found in a human overfeeding study (14), and we and our colleagues have demonstrated significant reductions in ghrelin levels accompanying weight gain from forced overfeeding in rats (Williams DL, Cummings DE, Kaplan JM and Grill HJ, unpublished results). Together, these observations suggest that ghrelin levels respond in a compensatory fashion to bidirectional alterations in body weight, consistent with the hypothesis that ghrelin contributes to the known adaptive metabolic responses to such alterations.

Because all of the conditions of weight loss cited above are characterized by low food intake, it is theoretically possible that the high ghrelin levels seen in them arise from decreased inhibitory input by ingested nutrients, rather than from weight loss *per se*. We have found, however, that weight loss achieved through a one-year aerobic exercise program without reduced caloric intake also triggers a rise in ghrelin levels in humans (53). Conversely, we found that mice subjected to mild caloric restriction, insufficient to cause weight loss (presumably because of compensatory changes in energy expenditure), show no change in plasma ghrelin levels (54). These observations suggest that ghrelin levels sense and respond to changes in body weight *per se* – not simply enteric nutrient load – although the mechanism for this type of ghrelin regulation is unknown.

4. ALTERATIONS OF CIRCULATING GHRELIN LEVELS IN PATHOPHYSIOLOGICAL CONDITIONS

Almost all reported perturbations of circulating ghrelin levels associated with disease states can be interpreted as adaptive responses of ghrelin to the body weight changes caused by these conditions.

4.1 Eating disorders

People with untreated anorexia nervosa are extremely underweight and manifest ghrelin levels that are among the highest reported in any humans (8-11,55). This observation indicates that impaired ghrelin secretion does not play an etiologic role in anorexia nervosa. The finding suggests a compensatory response of circulating ghrelin to energy deficit resulting from other causes that drive this psychiatric disorder. As would be predicted, elevated ghrelin levels among people with anorexia nervosa decrease with nutritional therapy, to a degree commensurate with the increase in body mass index (BMI), and levels can be fully normalized with adequate renutrition (9,11). Ghrelin concentrations in people with untreated anorexia nervosa are higher than those in equally lean individuals who are constitutionally thin (9). This observation suggests a response by ghrelin among anorectics to deviations below their innate body weight settling point. People with bulimia nervosa and normal BMI, who binge eat but do not purge, show normal ghrelin levels, as expected (55). Interestingly, one study reports that normal-weight bulimics who binge and purge have elevated ghrelin values – as high as those in patients with anorexia nervosa (55,56). These subjects' ghrelin levels correlated with the frequency of vomiting, and similarly, patients with anorexia nervosa who binged and purged had higher ghrelin values than did nonpurging anorectics (55). Elevated ghrelin levels among people with eating disorders who vomit frequently may result from an additional stimulatory effect, beyond weight loss, rendered by forced emptying of nutrients from the foregut – i.e., short-term ghrelin regulation. Alternatively, there may be a direct, perhaps autonomic, effect of vomiting itself on ghrelin.

4.2   Heart, liver and kidney failure

A study of patients with chronic congestive heart failure (CHF) found that affected individuals with cardiac cachexia had abnormally high ghrelin levels, whereas those without cachexia did not (50). Likewise, in a subset of CHF patients followed longitudinally for approximately one year, patients with cachexia demonstrated an increase in ghrelin levels, whereas those without cachexia had stable levels. Among all subjects, there was a progressive trend toward higher ghrelin levels with increasingly severe New York Heart Association functional status, but no relationship between ghrelin and left-ventricular ejection fraction, a direct measurement of cardiac performance itself. Together, these observations suggest an adaptive response of ghrelin to weight loss caused by heart failure, with no direct effect of cardiac function *per se* on ghrelin levels.

Similarly, patients with chronic liver disease have been reported to have increased serum ghrelin levels compared with healthy controls (51). Values were especially high among individuals with Child's Class C cirrhosis, independent of the etiology of liver failure. However, ghrelin levels were not related to measures of liver function itself – suggesting that most ghrelin is not cleared hepatically – but correlated significantly with clinical and biochemical markers of global illness and secondary complications. Thus, liver failure, like heart failure, probably increases ghrelin levels only indirectly by causing weight loss.

Patients with chronic renal failure also manifest elevated circulating ghrelin levels (52). These are progressively higher with increasing severity of disease, such that people with end-stage renal failure have ghrelin levels nearly three times higher than those in healthy controls. Unlike the situation with hepatic function, however, direct measures of renal function are significantly associated with ghrelin levels, which correlate positively with blood urea nitrogen (BUN) and creatinine, and negatively with creatinine clearance. A single course of hemodialysis decreases ghrelin levels by about one half – the same degree of clearance as occurs for BUN and creatinine. Conversely, bilateral nephrectomy in mice increases plasma ghrelin levels by more than three-fold within 16 hours (without affecting gastric ghrelin mRNA or protein levels), before significant weight loss occurs. These data provide compelling evidence that the kidney is a major site of clearance for ghrelin. Therefore, high circulating levels among patients with renal failure cannot be interpreted as a response to weight loss (although this may occur), because of the confounding effect of impaired ghrelin disposal. Interestingly, only the levels of total ghrelin (acylated and desacylated combined) correlate with measures of renal function, whereas those of acylated, bioactive ghrelin do not. Deacylation of ghrelin occurs very rapidly, mediated by putative circulating esterases, and the biological half-life of the acylated moiety is less than 10 minutes. This humoral degradation of bioactive ghrelin occurs so quickly that renal clearance may not have time to influence levels of acylated ghrelin.

4.3   A possible causal role for ghrelin in the obesity of Prader-Willi Syndrome

Ghrelin cannot be postulated to play a causal role in any of the disorders of body weight discussed above. Instead, alterations of circulating ghrelin levels in these conditions are all compatible with adaptive responses by ghrelin to changes in energy stores, consistent with the model of ghrelin as a barometer of such stores and a participant in long-term body weight regulation. Do pathological states exist wherein abnormal body weight is directly caused by disordered production of peripheral ghrelin? Clearly, this is not the case for common obesity. Because of the inverse relationship between BMI and circulating ghrelin levels, obese individuals have low ghrelin values, consistent with a compensatory, rather than causal, role for ghrelin in their overweight (6).

In an effort to identify putative rare people in whom obesity is driven by a primary overproduction of ghrelin, we reasoned that the phenotype of such individuals would resemble that of excessive signaling by NPY, one of the principal central targets of ghrelin action. Extensive studies in rodents suggest that constitutive activation of NPY should cause hyperphagia with consequent obesity, hypogonadotropic hypogonadism, and growth hormone (GH) dysregulation. These features of the predicted phenotype of excessive NPY, and thus of excessive ghrelin, are all characteristics of the PWS, a condition caused by imprinting defects in several genes at a locus on human chromosome 15 (57,58). Therefore, we examined ghrelin levels in adults with PWS and found them to be among the highest yet reported in humans: 4.5 times higher than those in equally obese controls without PWS, and 2.5 times higher than those in normal weight controls (7). Similar results have since been reported for a smaller but more extensively studied group of adults with PWS (59), as well as for children with the condition (12). The specificity of hyperghrelinemia in PWS is attested to by findings that people and rodents with other forms of genetic obesity – such as inactivating mutations of leptin, the leptin receptor, or the melanocortin-4 receptor – have ghrelin levels that are low and appropriate for their degree of adiposity (7,12). These findings also demonstrate that the high ghrelin levels characteristic of PWS are unlikely to arise as a secondary consequence of the functional leptin resistance seen in this condition, since the complete absence of leptin signaling in other diseases does not directly affect ghrelin levels. A theoretical cause of high ghrelin levels in people with PWS could be the caloric restriction that is imposed upon most affected individuals as part of their therapy. The extreme elevation of ghrelin levels in PWS, however, seems out of proportion to the degree of increase predicted from studies of severe caloric restriction in normal humans (46). Moreover, stimulation of ghrelin by calorie-restricted weight loss should produce higher levels among leaner individuals with PWS; yet we observed a complete loss of the normal correlation between ghrelin levels and BMI in these people (7). Extant data are consistent with a primary overproduction of ghrelin caused by the PWS genetic lesion. Although the genes encoding ghrelin and its receptor are not contained within the PWS locus, several of the affected genes in this condition encode transacting factors that could affect the expression of genes on other chromosomes (58).

Because the ghrelin levels observed in people with PWS are at least as high or higher than those shown to stimulate appetite and food intake when generated by peripheral ghrelin administration (26), hyperghrelinemia can be considered as a possible primary cause of the hyperphagia and obesity characteristic of PWS. Ghrelin may also contribute to the central hypogonadism and GH dysregulation of the syndrome, through excessive NPY signaling and possible override inhibition of GH, as has been reported with continuous GH-releasing hormone (GHRH) administration (60). There is no reason at this time to implicate ghrelin with regard to other features of

PWS, such as neonatal hypotonia, cognitive impairments, behavioral problems, or dysmorphic features (57).

If excessive ghrelin signaling indeed causes the hyperphagia and obesity of PWS, then people with this condition represent ideal candidates in whom to test the weight reducing efficacy of novel ghrelin-blocking agents. While awaiting the development of selective compounds of this sort, we investigated whether the high ghrelin levels in children with PWS could be lowered nonselectively by octreotide. This somatostatin analogue inhibits the secretion of many gastrointestinal hormones and suppresses ghrelin levels in normal humans (61). In a pilot study of children with PWS, we found that their high ghrelin levels were potently suppressed by a modest dose of octreotide administered for one week (62). This observation supports the feasibility of longer studies designed to assess the effect of such treatment on appetite, body weight, and other relevant parameters. Even if such trials are successful, however, it is likely that more specific anti-ghrelin agents that do not affect other hormones would be better ultimate treatment options for this condition.

5. METHODS OF WEIGHT LOSS THAT DO NOT INCREASE CIRCULATING GHRELIN LEVELS: LOW-FAT DIETS AND GASTRIC BYPASS SURGERY

Because an increase in peripheral ghrelin levels may be one of the adaptive responses that defends against weight loss, we investigated the effects on ghrelin of weight loss achieved by two traditionally successful methods: low-fat diets and gastric bypass surgery.

Although dietary restriction in general is notoriously ineffective at durably maintaining substantial weight loss (13,63), low-fat diets have been extensively shown to promote modest weight loss in multiple species (64). Because of the relative efficacy of low-fat diet-induced weight loss, we examined the impact of this intervention on ghrelin levels. Human subjects who lost 5.1% of their body weight through *ad-libitum* consumption of a 15% fat diet for three months, failed to show any alteration in 24-hour plasma ghrelin profiles (65). This finding contrasts with the increases in ghrelin levels that are triggered by multiple other modes of weight loss, enumerated above. The lack of change in ghrelin levels in our study cannot be explained by an insufficient magnitude of weight loss among these subjects, since individuals who lost lesser amounts of weight with either chronic exercise or mixed lifestyle modifications showed significantly increased ghrelin levels (48,53). Furthermore, we found no increase in ghrelin levels among mice that lost 17% of body weight following consumption of a very low fat (10%) diet for one month. In contrast, mice that lost comparable weight because of a low calorie diet manifested a 70% increase in ghrelin levels (54). We do not yet know why low-fat diet-induced weight loss fails to elicit the normal compensatory ghrelin response. It is possible, however, that the absence of this adaptive response contributes to the relative success of low fat diets in promoting weight loss. It will interesting to determine the effect on ghrelin levels of weight loss resulting from the currently popular low carbohydrate diets, which have recently been validated as being moderately effective methods to lose weight, at least temporarily (reviewed in Ref. 66).

Roux-en-Y gastric bypass (RYGB) surgery is the most effective approved method of weight loss currently available (67,68). The procedure restricts the gastric volume that is capable of storing food, bypassing most of the stomach and all of the duodenum with a gastro-jejunal anastomosis. RYGB causes 35-40% loss of body weight, most of which is maintained in follow-up studies lasting up to 20 years (68-70; Pories WJ, personal communication). The mechanisms underlying this dramatic effect are enigmatic (reviewed in Ref. 5). Unquestionably, restriction of the gastric volume capable of accommodating ingested food causes early satiety, with consequent reductions in meal size. Post-operative patients, however, typically describe a generalized loss of hunger throughout the day – not just after meals. Accordingly, they eat fewer (as well as smaller) meals per day, and voluntarily restrict consumption of high-calorie foods and beverages. Presumably because of these changes, weight loss following RYGB is greater than that from vertical-banded gastroplasty, even though the latter procedure imposes at least as much gastric restriction. It is unlikely that the difference in efficacy between RYGB and gastroplasty results from either RYGB-induced malabsorption (which is not clinically significant after the standard, proximal procedure) or dumping syndrome (which develops inconsistently). The unexplained changes in eating behavior and profound loss of appetite that result from RYGB suggest that anorexigenic alterations beyond simple gastric restriction occur after this operation, contributing to weight loss.

Because most ghrelin production occurs in the anatomic region affected by RYGB, we investigated the effect of this operation on circulating ghrelin. In a small study, we found that ghrelin levels in patients who had undergone RYGB 1.4±0.4 years earlier were markedly lower than those in either normal weight or matched-obese control subjects, and showed neither the prandial oscillations nor diurnal rhythm that characterize normal 24-hour ghrelin profiles (46). These low ghrelin values were especially remarkable because the post-RYGB patients had lost 36±5% of their body weight – a change that would have triggered an increase in ghrelin levels if it had been achieved by most other means (see above).

Three other groups have subsequently made similar observations (71-73). These investigators found either unexpectedly low ghrelin levels in post-RYGB subjects compared with appropriate controls, or a fall in ghrelin levels after surgery among patients followed prospectively. Another group reported ghrelin levels that were stable after RYGB (74). This lack of change in ghrelin was interpreted as inappropriate, in view of the massive (36%) weight loss that had occurred. Finally, a Swedish group observed that weight loss following their version of RYGB was accompanied by increased ghrelin levels, as would occur with most other mechanisms of weight loss (75). Together, these findings suggest that, through cryptic mechanisms, RYGB usually suppresses ghrelin levels, and this effect may contribute to the weight-reducing impact of the procedure. However, ghrelin suppression after RYGB is not universal. Important questions for future studies include whether the decrease in ghrelin levels, when it occurs, augments surgical weight loss, and if so, what methodological details yield this result in the hands of some surgeons but not others (see Ref. 5 and 76 for more details).

We initially speculated that ghrelin levels fall after RYGB because of a phenomenon called "override inhibition" (46). According to this model, isolating the majority of ghrelin-producing tissue from direct contact with enteral nutrients – an intervention that would acutely stimulate ghrelin production – paradoxically inhibits it when present permanently after RYGB. This process would resemble the paradoxical suppression of gonadotropins or GH by continuous signaling from the stimulatory peptides, gonadotropin-releasing hormone or GHRH, respectively (60,77). The override inhibition hypothesis predicts that purely restrictive bariatric procedures that do not bypass any of the gastrointestinal system should not suppress ghrelin. Consistent with this prediction, we found that weight loss achieved with adjustable gastric banding was accompanied by a durable increase in plasma ghrelin levels, as would be expected from other modes of weight loss (78). It is possible that differential effects on ghrelin could help explain the greater weight-reducing efficacy of RYGB compared with purely restrictive procedures, such as gastroplasty (reviewed in Ref. 5).

Another possible mechanism to explain sustained ghrelin suppression following RYGB is denervation of the sympathetic and/or parasympathetic input to the foregut – effects that are variably wrought by different bariatric surgeons (76). In collaboration with the laboratory of Dr. Harvey Grill, we have found that rat ghrelin is not regulated by direct contact between nutrients in the gut lumen and ghrelin cells in either the stomach or duodenum (the principal sites of production) (30). We also found that vagotomy disrupts normal ghrelin regulation (79). Based on these observations, we now favor an autonomic explanation, rather than override inhibition, to account for the RYGB findings. Additional studies, however, are required to clarify the mechanism by which this operation sometimes suppresses ghrelin production, and therefore, to identify the details of surgical technique that should be followed in order to seek the effect expressly.

If definitive, interventional studies confirm that ghrelin suppression contributes to weight loss after RYGB, this finding would imply that pharmacological blockade of ghrelin signaling may achieve some of the profound weight loss that results from RYGB. Several pharmaceutical companies are vigorously developing ghrelin receptor antagonists and inverse agonists, and orally bioavailable ghrelin agonists have already been studied as GH secretagogues in human trials. The next few years promise to be an exciting time for ghrelin research, as we learn whether these reagents can be used to combat obesity and/or wasting disorders.

REFERENCES

1. Cummings DE, Purnell JQ, Frayo RS, Schmidova K, Wisse BE, Weigle DS. A preprandial rise in plasma ghrelin levels suggests a role in meal initiation in humans. Diabetes. 2001; 50:1714-9.
2. Tschop M, Wawarta R, Riepl RL, et al. Post-prandial decrease of circulating human ghrelin levels. J Endocrinol Invest. 2001; 24:RC19-21.
3. Sugino T, Yamaura J, Yamagishi M, et al. A transient surge of ghrelin secretion before feeding is modified by different feeding regimens in sheep. Biochem Biophys Res Commun. 2002; 298:785-8.
4. Tolle V, Bassant M-H, Zizzari P, et al. Ultradian rhythmicity of ghrelin secretion in relation with GH, feeding behavior, and sleep-wake patterns in rats. Endocrinology. 2002; 143:1353-61.
5. Cummings DE, Shannon MH. Roles for ghrelin in the regulation of appetite and body weight. Arch Surg. 2003; 138:389-96.
6. Tschop M, Weyer C, Tataranni PA, Devanarayan V, Ravussin E, Heiman ML. Circulating ghrelin levels are decreased in human obesity. Diabetes. 2001; 50:707-9.
7. Cummings DE, Clement K, Purnell JQ, et al. Elevated plasma ghrelin levels in Prader-Willi syndrome. Nat Med. 2002; 8:643-4.
8. Ariyasu H, Takaya K, Tagami T, et al. Stomach is a major source of circulating ghrelin, and feeding state determines plasma ghrelin-like immunoreactivity levels in humans. J Clin Endocrinol Metab. 2001; 86:4753-8.
9. Tolle V, Kadem M, Bluet-Pajot MT, et al. Balance in ghrelin and leptin levels in anorexia nervosa patients and constitutionally thin women. J Clin Endocrinol Metab. 2003; 88:109-16.
10. Shiiya T, Nakazato M, Mizuta M, et al. Plasma ghrelin levels in lean and obese humans and effect of glucose on ghrelin secretion. J Clin Endocrinol Metab. 2002; 87:240-4.
11. Otto B, Cuntz U, Fruehauf E, et al. Weight gain decreases elevated plasma ghrelin concentrations of patients with anorexia nervosa. Eur J Endocrinol. 2001; 145:669-73.

12. Haqq AM, Farooqi S, O'Rahilly S, et al. Serum ghrelin levels are inversely correlated with body mass index, age, and insulin concentrations in normal children and are markedly increased in Prader-Willi syndrome. J Clin Endocrinol Metab. 2003; 88:174-8.

13. Cummings DE, Schwartz MW. Genetics and pathophysiology of human obesity. Annu Rev Med. 2003; 54:453-71.

14. Ravussin E, Tschop M, Morales S, Bouchard C, Heiman ML. Plasma ghrelin concentration and energy balance: overfeeding and negative energy balance studies in twins. J Clin Endocrinol Metab. 2001; 86:4547-51.

15. Kojima M, Hosoda H, Date Y, Nakazato M, Matsuo H, Kangawa K. Ghrelin is a growth-hormone-releasing acylated peptide from stomach. Nature. 1999; 402:656-60.

16. Date Y, Kojima M, Hosoda H, et al. Ghrelin, a novel growth hormone-releasing acylated peptide, is synthesized in a distinct endocrine cell type in the gastrointestinal tracts of rats and humans. Endocrinology. 2000; 141:4255-61.

17. Gnanapavan S, Kola B, Bustin SA, et al. The tissue distribution of the mRNA of ghrelin and subtypes of its receptor, GHS-R, in humans. J Clin Endocrinol Metab. 2002; 87:2988-91.

18. Krsek M, Rosicka M, Haluzik M, et al. Plasma ghrelin levels in patients with short-bowel syndrome. Endocrine Research. 2002; 28:27-33.

19. Tschop M, Smiley DL, Heiman ML. Ghrelin induces adiposity in rodents. Nature. 2000; 407:908-13.

20. Asakawa A, Inui A, Kaga T, et al. Ghrelin is an appetite-stimulatory signal from stomach with structural resemblance to motilin. Gastroenterology. 2001; 120:337-45.

21. Nakazato M, Murakami N, Date Y, et al. A role for ghrelin in the central regulation of feeding. Nature. 2001; 409:194-8.

22. Wren AM, Small CJ, Ward HL, et al. The novel hypothalamic peptide ghrelin stimulates food intake and growth hormone secretion. Endocrinology. 2000; 141:4325-8.

23. Faulconbridge LF, Cummings DE, Kaplan JM, Grill HJ. Hyperphagic effects of brainstem ghrelin administration. Diabetes. 2003; 52:2260-5.

24. Masuda Y, Tanaka T, Inomata N, et al. Ghrelin stimulates gastric acid secretion and motility in rats. Biochem Biophys Res Commun. 2000; 276:905-8.

25. Lu XY, Shieh KR, Kabbaj M, Barsh GS, Akil H, Watson SJ. Diurnal rhythm of agouti-related protein and its relation to corticosterone and food intake. Endocrinology. 2002; 143:3905-15.

26. Wren AM, Seal LJ, Cohen MA, et al. Ghrelin enhances appetite and increases food intake in humans. J Clin Endocrinol Metab. 2001; 86:5992-5.

27. Wren AM, Small CJ, Abbott CR, et al. Ghrelin causes hyperphagia and obesity in rats. Diabetes. 2001; 50:2540-7.

28. Cummings DE, Frayo RS, Louis-Sylvestre J, Chapelot D. Human plasma ghrelin levels are temporally linked to the perception of hunger, and spontaneous preprandial ghrelin surges predict voluntary meal initiation. Program of the 85th Annual Meeting of The Endocrine Society, San Francisco, CA, 2002; OR34-2 (abstract).

29. Callahan HS, Cummings DE, Pepe MS, Breen PA, Matthys CC, Weigle DS. Postprandial suppression of plasma ghrelin level is proportional to ingested caloric load in humans. 63rd Annual American Diabetes Association Meeting, New Orleans, LA, 2003 (abstract).

30. Williams DA, Cummings DE, Grill HJ, Kaplan JM. Meal-related ghrelin suppression requires postgastric feedback. Endocrinology. 2003; 144:2765-7.

31. Schwartz MW, Woods SC, Porte D, Seeley RJ, Baskin DG. Central nervous system control of food intake. Nature. 2000; 404:661-71.

32. Banks WA, Tschop M, Robinson SM, Heiman ML. Extent and direction of ghrelin transport across the blood-brain barrier is determined by its unique primary structure. J Pharmacol Exp Ther. 2002; 302:822-7.

33. Willesen MG, Kristensen P, Romer J. Co-localization of growth hormone secretagogue receptor and NPY mRNA in the arcuate nucleus of the rat. Neuroendocrinology. 1999; 70:306-16.

34. Cowley MA, Smith RG, Diano S, et al. The distribution and mechanism of action of ghrelin in the CNS demonstrate a novel hypothalamic circuit regulating energy homeostasis. Neuron. 2003; 37:649-61.

35. Dickson SL, Luckman SM. Induction of *c-fos* messenger ribonucleic acid in neuropeptide Y and growth hormone (GH)-releasing factor neurons in the rat arcuate nucleus following systemic injection of the GH secretagogue, GH-releasing peptide-6. Endocrinology. 1997; 138:771-7.

36. Kamegai J, Tamura H, Shimizu T, Ishii S, Sugihara H, Wakabayashi I. Central effect of ghrelin, an endogenous growth hormone secretagogue, on hypothalamic peptide gene expression. Endocrinology. 2000; 141:4797-800.

37. Kamegai J, Tamura H, Shimizu T, Ishii S, Sugihara H, Wakabayashi I. Chronic central infusion of ghrelin increases hypothalamic neuropeptide Y and agouti-related protein mRNA levels and body weight in rats. Diabetes. 2001; 50:2438-43.

38. Shintani M, Ogawa Y, Ebihara K, et al. Ghrelin, an endogenous growth hormone secretagogue, is a novel orexigenic peptide that antagonizes leptin action through the activation of hypothalamic neuropeptide Y/Y1 receptor pathway. Diabetes. 2001; 50:227-32.

39. Matsumura K, Tsuchihashi T, Fujii K, Abe I, Iida M. Central ghrelin modulates sympathetic activity in conscious rabbits. Hypertension. 2002; 40:694-9.

40. Lawrence CB, Snape AC, Baudoin FM, Luckman SM. Acute central ghrelin and GH secretagogues induce feeding and activate brain appetite centers. Endocrinology. 2002; 143:155-62.

41. Shuto Y, Shibasaki T, Otagiri A, et al. Hypothalamic growth hormone secretagogue receptor regulates growth hormone secretion, feeding, and adiposity. J Clin Invest. 2002; 109:1429-36.

42. Murakami N, Hayashida T, Kuroiwa T, et al. Role for central ghrelin in food intake and secretion profile of stomach ghrelin in rats. J Endocrinol. 2002; 174:283-8.

43. Bagnasco M, Tulipano G, Melis MR, Argiolas A, Cocchi D, Muller EE. Endogenous ghrelin is an orexigenic peptide acting in the arcuate nucleus in response to fasting. Regul Pept. 2003; 111:161-7.

44. Asakawa A, Inui A, Kaga T, et al. Antagonism of ghrelin receptor reduces food intake and body weight gain in mice. Gut. 2003; 52:947-52.

45. Ukkola O, Ravussin E, Jacobson P, et al. Role of ghrelin polymorphisms in obesity based on three different studies. Obes Res. 2002; 10:782-91.

46. Cummings DE, Weigle DS, Frayo RS, et al. Human plasma ghrelin levels after diet-induced weight loss and gastric bypass surgery. New Engl J Med. 2002; 346:1623-30.

47. Wisse BE, Frayo RS, Schwartz MW, Cummings DE. Reversal of cancer anorexia by blockade of central melanocortin receptors in rats. Endocrinology. 2001; 142:3292-301.

48. Hansen TK, Dall R, Hosoda H, et al. Weight loss increases circulating levels of ghrelin in human obesity. Clin Endocrinol (Oxf). 2002; 56:203-6.

49. Shimizu Y, Noritoshi N, Isobe T, et al. Increased plasma ghrelin levels in lung cancer cachexia. Clin Cancer Res. 2003; 9:774-8.

50. Nagaya N, Uematsu M, Kojima M, et al. Elevated circulating level of ghrelin in cachexia associated with chronic heart failure. Circulation. 2001; 104:2034-8.

51. Tacke F, Brabant G, Kruck E, et al. Ghrelin in chronic liver disease. J Hepatol. 2003; 38:447-54.

52. Yoshimoto A, Mori K, Sugawara A, et al. Plasma ghrelin and desacyl ghrelin concentrations in renal failure. J Am Soc Nephrol. 2002; 13:2748-52.

53. Foster KE, McTiernan A, Frayo RS, Rajan B, Cummings DE. Changes in human plasma ghrelin and leptin levels correlate negatively with each other during aerobic exercise training. Program of the 84th Annual Meeting of The Endocrine Society, Philadelphia, PA, 2002; P3-438 (abstract).

54. Bush EN, Droz B, Shapiro R, et al. Effects of low-calorie or low-fat diet-induced weight loss on plasma ghrelin levels. Program of the 85th Annual Meeting of the Endocrine Society, Philadelphia, PA, 2003; P3-97 (abstract).

55. Tanaka M, Naruo T, Nagai N, et al. Habitual binge-purge behavior influences circulating ghrelin levels in eating disorders. J Psych Res. 2003; 37:17-22.

56. Tanaka M, Naruo T, Muranaga T, et al. Increased fasting plasma ghrelin levels in patients with bulimia nervosa. Eur J Endocrinol. 2002; 146:R1-3.

57. Burman P, Ritzen EM, Lindgren AC. Endocrine dysfunction in Prader-Willi syndrome: a review with special reference to GH. Endocr Rev. 2001; 22:787-99.

58. Nicholls RD, Knepper JL. Genome organization, function, and imprinting in Prader-Willi and Angelman syndromes. Annu Rev Genomics Hum Genet. 2001; 2:153-75.

59. DelParigi A, Tschop M, Heiman ML, et al. High circulating ghrelin: a potential cause for hyperphagia and obesity in Prader-Willi syndrome. J Clin Endocrinol Metab. 2002; 87:5461-4.

60. Rittmaster RS, Loriaux DL, Merriam GR. Effect of continuous somatostatin and growth hormone-releasing hormone (GHRH) infusions on the subsequent growth hormone (GH) response to GHRH: Evidence for somatotroph desensitization independent of GH pool depletion. Neuroendocrinology. 1987; 45:118-22.

61. Norrelund H, Hansen TK, Orskov H, et al. Ghrelin immunoreactivity in human plasma is suppressed by somatostatin. Clin Endocrinol (Oxf). 2002; 57:539-46.

62. Haqq AM, Stadler DD, Rosenfeld RG, et al. Circulating ghrelin levels are suppressed by meals and octreotide therapy in children with Prader-Willi syndrome. J Clin Endocrinol Metab. 2003; 88:3573-6.

63. Yanovski SZ, Yanovski JA. Obesity. N Engl J Med. 2002; 346:591-602.

64. Astrup A, Astrup A, Buemann B, Flint A, Raben A. Low-fat diets and energy balance: how does the evidence stand in 2002. Proc Nutr Soc. 2002; 61:299-309.

65. Weigle DS, Cummings DE, Newby PD, et al. Roles of leptin and ghrelin in the loss of body weight caused by low-fat, high-carbohydrate diet. J Clin Endocrinol Metab. 2003; 88:1577-86.

66. Bonow RO, Eckel RH. Diet, obesity, and cardiovascular risk. N Engl J Med. 2003; 348:2057-8.

67. Consensus DCP. Gastrointestinal surgery for severe obesity. Ann Intern Med. 1991; 115:956-61.

68. Mun EC, Blackburn GL, Matthews JB. Current status of medical and surgical therapy for obesity. Gastroenterology. 2001; 120:669-81.

69. Brolin RE. Bariatric surgery and long-term control of morbid obesity. JAMA. 2002; 288:2793-6.

70. Pories WJ, Swanson MS, MacDonald KG, et al. Who would have thought it? An operation proves to be the most effective therapy for adult-onset diabetes mellitus. Ann Surg. 1995; 222:339-52.

71. Geloneze B, Tambascia MA, Pilla VF, Geloneze SR, Repetto EM, Pareja JC. Ghrelin: a gut-brain hormone: effect of gastric bypass surgery. Obesity Surgery. 2003; 13:17-22.

72. Tritos NA, Mun E, Bertkau A, Grayson R, Maratos-Flier E, Goldfine A. Serum ghrelin levels in response to glucose load in obese subjects post-gastric bypass surgery. Obes Res. 2003; 11:919-24.

73. Fruhbeck G, Diez-Caballero A, Gil MJ, et al. Fundus functionality and ghrelin concentrations after bariatric surgery. N Engl J Med. 2004; 350:308-9.

74. Faraj M, Havel PJ, Phelis S, Blank D, Sniderman AD, Cianflone K. Plasma acylation-stimulating protein, adiponectin, leptin, and ghrelin before and after weight loss induced by gastric bypass surgery in morbidly obese subjects. J Clin Endocrinol Metab. 2003; 88:1594-602.

75. Holdstock C, Engstrom BE, Obrvall M, Lind L, Sundborn M, Karlsson FA. Ghrelin and adipose tissue regulatory peptides: effect of gastric bypass surgery in obese humans. J Clin Endocrinol Metab. 2003; 88:3177-83.

76. Cummings DE, Shannon MH. Ghrelin and gastric bypass: Is there a hormonal contribution to surgical weight loss? J Clin Endocrinol Metab. 2003; 88:2999-3002.

77. Belchetz PE, Plant TM, Nakai Y, Keogh EJ, Knobil E. Hypophysial responses to continuous and intermittent delivery of hypothalamic gonadotropin-releasing hormone. Science. 1978; 202:631-3.

78. Cummings DE, Coupaye M, Frayo RS, Guy-Grand B, Basdevant A, Clement K. Weight loss caused by adjustable gastric banding increases plasma ghrelin levels in humans. Program of the 85th Annual Meeting of The Endocrine Society, Philadelphia, PA, 2003; OR33-3 (abstract).

79. Williams DL, Grill HJ, Cummings DE, Kaplan JM. Vagotomy dissociates short- and long-term controls of circulating ghrelin. Endocrinology. 2003; 144:5184-7.

Chapter 15

GHRELIN MEASUREMENT: PRESENT AND PERSPECTIVES

Hiroshi Hosoda & Kenji Kangawa

Department of Biochemistry, National Cardiovascular Center Research Institute, Suita, Osaka 565-8565 and Translational Research Center, Kyoto University Hospital, Kyoto, 606-8507, Japan

Abstract: Ghrelin, a 28-amino acid peptide with an *n*-octanoyl modification indispensable for its biological activity, is synthesized principally in the stomach and released in response to acute and chronic energy imbalances. Due to increased interest in ghrelin measurement, a standardized method of sample collection is required. The present study sought to investigate the effect of a variety of anticoagulants and storage conditions on ghrelin stability. In whole blood and plasma, acylated ghrelin was found to be highly unstable, as ester bonding can be degraded both chemically and enzymatically under these conditions. To acquire accurate data on ghrelin levels, blood samples should be collected with ethylene diamine tetra-acetic acid-aprotinin and centrifuged within 30 minutes under cooled conditions. The stability of ghrelin gradually decreased in untreated plasma, with 40% degradation after 6 hours of storage at 37°C. Storage at 4°C and acidification of plasma to pH 3-4 maintained ghrelin stability. In acidified plasma (pH 4), ghrelin levels were not altered by up to 4 cycles of freezing and thawing. After intravenous administration to anesthetized rats, plasma ghrelin levels rapidly decreased with a half-life of 8 minutes. In conclusion, as ghrelin is highly unstable, it is necessary to standardize the preparation of samples to ensure reliable ghrelin measurements.

Key words: assay, RIA, pharmacokinetic

1. INTRODUCTION

Ghrelin is an acylated peptide with growth hormone (GH) releasing activity (1,2). It was first isolated from rat and human stomachs during the search for an endogenous ligand to the "orphan" G-protein coupled receptor, GH Secretagogue (GHS) receptor (GHS-R) (3-5). The peptide contains 28-amino acids, with *n*-octanoylation of the serine 3 hydroxyl group necessary for biological activity. Attempts to understand the post-translational mechanism of octanoyl modification, important for the regulation of ghrelin biological activity, opens up a new field in protein chemistry as octanoylation has not been previously observed as a regulable peptide modification. The concentration of circulating ghrelin, an important regulator of GH secretion and energy homeostasis (6-11), is influenced by acute and chronic changes in nutritional state (12-17). Most studies focused on the somatotrophic and orexigenic role of ghrelin; little is known, however, about the kinetics of ghrelin expression and regulation. As ester bonding is both chemically and enzymatically unstable, deletion of the octanoyl modification of ghrelin can occur during the storage, handling and/or dissolution in culture medium (18). Thus, "active" ghrelin could be highly unstable in blood samples. We established two radioimmunoassays (RIAs) specifically sensitive to detect ghrelin: one recognizes only the active, octanoyl-modified ghrelin and the other, recognizing the C-terminal portion of ghrelin, can be used to measure total ghrelin (19,20). Focusing on the active form of ghrelin, we compared the reliability of ghrelin measurements in various plasma and serum samples to evaluate data on the stability of ghrelin under different storage conditions.

2.   MATERIALS AND METHODS

2.1   Materials

Rat and human ghrelin synthesis was performed as previously described (1). Briefly, fully-protected rat and human ghrelin (with the exception of the exposed hydroxyl group of serine 3) was synthesized by the Fmoc solid-phase method using a peptide synthesizer (433A, Applied Biosystems, Foster City, CA). The hydroxyl group of serine 3 was then acylated with *n*-octanoic acid by the action of 1-ethyl-3-(3-dimethylaminopropyl) carbodiimide in the presence of 4-(dimethylamino) pyridine. Synthesized peptides were purified by reverse-phase high performance liquid chromatography (RP-HPLC).

2.2   RIA for ghrelin

The two RIAs measuring plasma ghrelin were performed as described (19). In brief, plasma ghrelin levels were measured with two RIAs using two polyclonal rabbit antibodies raised against either the N-terminal [1-11] (Gly1-Lys11) or C-terminal [13-28] (Gln13-Arg28) fragments of rat ghrelin. RIA incubation mixtures, containing 100 µl of either standard ghrelin or diluted plasma/serum sample with 200 µl of antiserum diluted in RIA buffer containing 0.5% normal rabbit serum, were initially incubated for 12 hours. Next, 100 µl of ^{125}I-labeled tracer (15,000 cpm) was added for a 36 hours incubation. Anti-rabbit IgG goat serum (100 µl) was added prior to an additional 24 hours incubation. Free and bound tracers were then separated by centrifugation at 3,000 rpm for 30 minutes. Following aspiration of the supernatant, radioactivity in the pellet was quantitated using a gamma counter (ARC-600, Aloka, Tokyo). All assays were performed in duplicate at 4°C. The anti-rat ghrelin[1-11] antiserum (#G606) specifically recognizes rat n-octanoylated ghrelin and does not detect des-acyl ghrelin. The anti-rat ghrelin[13-28] antiserum (#G107) equally recognizes both the n-octanoyl modified and des-acyl rat ghrelin forms. Both antisera were equally cross-reactive with human ghrelin. We did not observe significant cross-reactivity with other peptides. In the following sections, the RIA procedures using G606 antiserum is termed N-RIA, while that utilizing the G107 antiserum is termed C-RIA. The respective intra- and inter-assay coefficients of variation were 3% and 6% for the N-RIA and 6% and 9% for the C-RIA.

2.3   Study protocol

2.3.1   Study 1

This study sought to compare the reliability of ghrelin measurements in serum and plasma samples. All blood samples were taken from healthy volunteers (n=3) who were not currently taking medication. Blood was taken from the forearm vein and immediately divided into tubes for serum and plasma preparation using either ethylene diamine tetra-acetic acid (EDTA)-2Na (1 mg/ml) with aprotinin (500 kIU/ml), EDTA-2Na, or heparin sodium as anticoagulants. Synthetic human ghrelin was added to each blood sample at a final concentration of 40 ng/ml, then sequentially divided into two aliquots for incubation at either 37°C or 4°C. After 0, 30 and 60 minutes incubations, blood samples were centrifuged, diluted, and subjected to ghrelin-specific RIAs.

2.3.2   Study 2

We examined the effect of plasma pH on ghrelin stability. The EDTA-aprotinin-treated plasma was divided into five samples; the pH was then adjusted to 3, 4, 5, 6, or 7.4 with 1 N hydrogen chloride (HCl). Synthetic human ghrelin was then added to each EDTA-aprotinin plasma aliquot at a final concentration of 75 ng/ml. Then, each of the five plasma aliquots was subdivided into two aliquots and stored at either 4°C or 37°C. Plasma testing was repeated at 0, 1, 2, 4 and 6 hours for samples stored at 37°C or 0, 3 and 6 hours for samples incubated at 4°C.

2.3.3   Study 3

We evaluated the effect of repeated freezing and thawing on the stability of ghrelin. EDTA-aprotinin-treated plasma samples were divided into two pH groups, one of which was acidified with 1 N HCl to pH 4, while the other was not acidified (pH 7.4). Following the addition of synthetic human ghrelin (75 ng/ml), we subjected samples to four freezing and thawing cycles. Test samples were obtained before freezing and after rethawing; the recovery of human ghrelin was then assessed by ghrelin-specific RIAs.

2.3.4   Study 4

We assessed the kinetics of rat ghrelin plasma concentrations after intravenous administration to anesthetized rats. Male Wistar rats weighing 300-350 g (Chales River Japan, Yokohama, Japan) were anesthetized with pentobarbital (50 mg/kg, intraperitoneally) and canulated in the jugular artery and vein for blood sampling and infusion, respectively. Synthetic rat ghrelin (100 μg/body) was administrated intravenously. Blood (0.2 ml) was drawn prior to infusion and at 1, 3, 5, 7, 10, 15, 20, 30, 45, 60, 75 and 90 minutes after peptide administration. Each blood sample was collected in a polypropylene tube containing EDTA-2Na (2 mg/ml) with aprotinin (500 kIU/ml). Plasma was immediately separated by centrifugation. Plasma half-life of ghrelin was calculated using WinNonlin Standard program (Ver. 3.1) (2-compartment model, Scientific Consulting Inc.). All procedures were performed in accordance with the Japanese Physiological Society's guidelines for animal care.

3. RESULTS

3.1 RIA for ghrelin

Plasma ghrelin levels were measured using two ghrelin-specific RIA systems, N-RIA recognizing active, acylated ghrelin and C-RIA detecting total ghrelin (Figure 15-1). As rat and human ghrelin possess only two amino acid differences, these antisera exhibited good cross-reactivity between rat and human ghrelins.

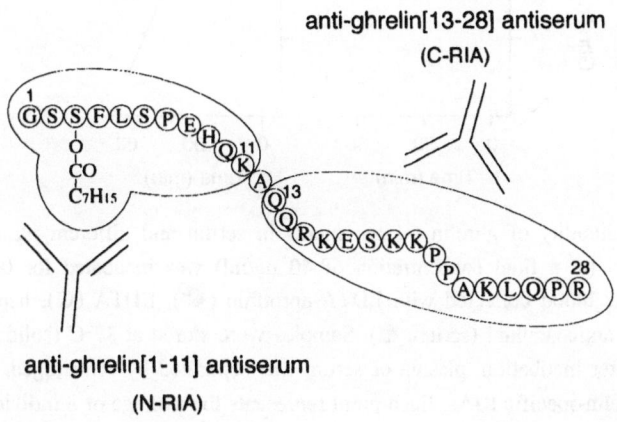

Figure 15-1. The two ghrelin-specific RIA systems: N-RIA recognize acylated ghrelin and C-RIA detect total ghrelin.

3.2 Reliability of ghrelin measurements in serum and plasma samples

The effect of different anticoagulants on the detected ghrelin levels is compared in Figure 15-2. Although the serum and three different plasma samples tested gave comparable results for total ghrelin levels by C-RIA, ghrelin levels in these samples by N-RIA decreased significantly at 37°C. Serum samples were highly affected by such treatment; samples stored for 60 minutes at 37°C lost approximately 35% of ghrelin in comparison with the basal levels at 0 minutes. The recovery of ghrelin following the use of heparin-plasma as an anticoagulant was also decreased. Use of EDTA-aprotinin for plasma treatment gave lower decreases in ghrelin stability than other procedures. Storage at 4°C also improved ghrelin stability, making this treatment sufficient for sampling within 30 minutes of sample isolation.

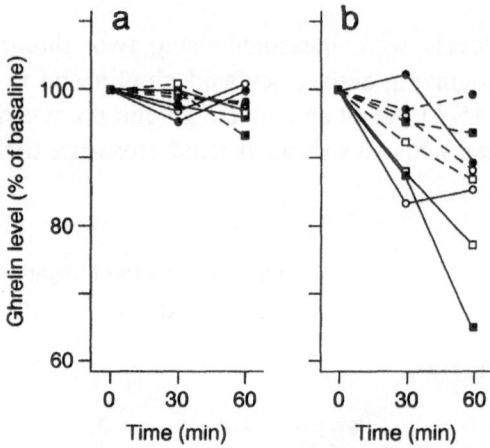

Figure 15-2. Reliability of ghrelin measurements in serum and different plasma samples. Synthetic ghrelin (at a final concentration of 40 ng/ml) was incubated for 0, 30 and 60 minutes in whole blood collected with EDTA-aprotinin (●), EDTA (O), heparin sodium (□), or without anticoagulant (serum, ■). Samples were stored at 37°C (solid line) or 4°C (dotted line). After incubation, plasma or serum was separated by centrifuged, diluted, and subjected to ghrelin-specific RIAs. Each point represents the average of 3 individual samples assessed by C-RIA (a) and N-RIA (b).

3.3   Stability of ghrelin in plasma storage conditions

As ester bonds are often unstable in biological materials, we investigated the stability of ghrelin in human EDTA-aprotinin-treated plasma. The effect of acidification on ghrelin stability in plasma is summarized in Figure 15-3. When stored at 37°C, ghrelin levels gradually decreased at all the pH values tested, however, ghrelin was most stable in highly acidified plasma samples of pH 3-4. At a plasma pH between 3-4 and a storage temperature of 4°C, the stability of ghrelin did not change significantly over a 6-hour period. By C-RIA, ghrelin levels remained stable across the different pH and storage temperature conditions.

Repeated freezing and thawing also influenced on the ghrelin stability (Figure 15-4). Ghrelin levels in untreated plasma samples decreased with each successive cycle, whereas ghrelin stability remained relatively constant following acidification. Ghrelin levels by C-RIA were unchanged despite repeated freeze-thaw treatments in both acidified and untreated plasma samples.

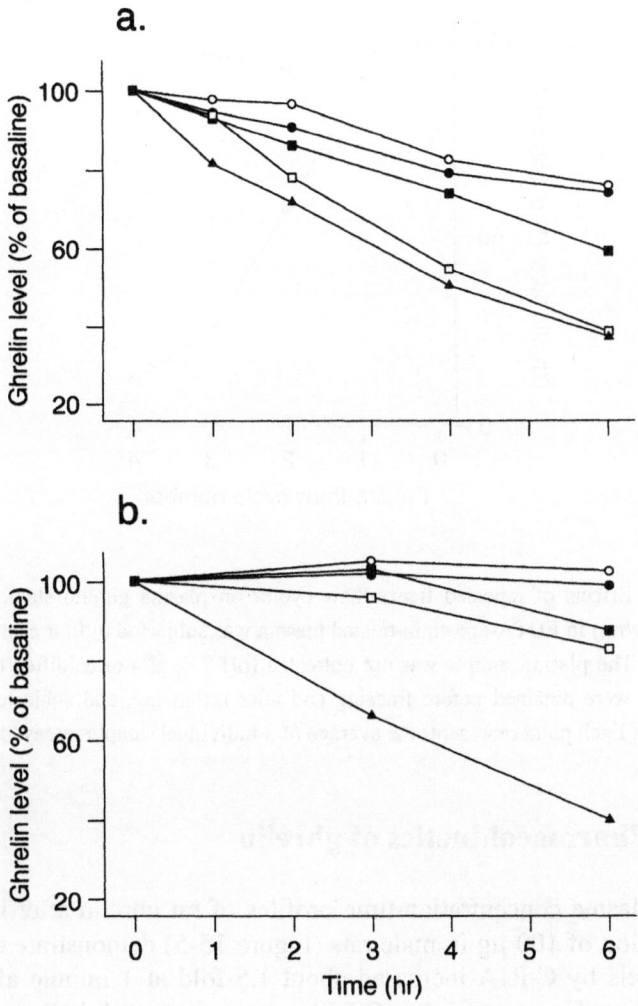

Figure 15-3. Effects of storage pH, duration and temperature on plasma ghrelin stability. Synthetic ghrelin (at a final concentration of 75 ng/ml) was incubated in EDTA-aprotinin-treated plasma at various pH values (pH 7.4, ▲; pH 6, □; pH 5, ■; pH 4, ●; pH 3, ○). After incubation, plasma sample was diluted, and subjected to RIAs for ghrelin. Each point represents the average of 3 individual samples assayed by N-RIA for 37°C (a) and 4°C (b).

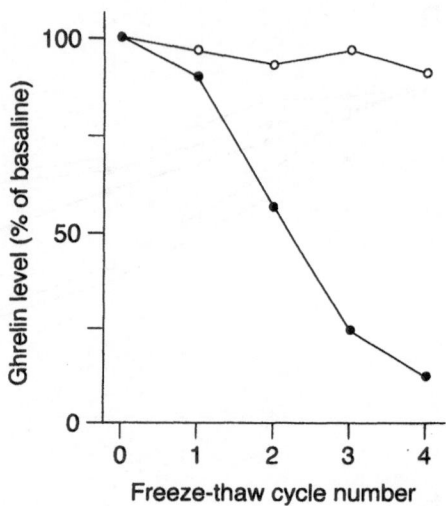

Figure 15-4. Effects of repeated freeze-thaw cycles on plasma ghrelin stability. Synthetic ghrelin (75 ng/ml) in EDTA-aprotinin-treated plasma was subjected to four cycles of freezing and thawing. The plasma sample was not untreated (pH 7.4, ●) or acidified to pH 4 (○). Test samples were obtained before freezing and after rethawing, and subjected to ghrelin-specific RIAs. Each point represents the average of 3 individual samples assayed by N-RIA.

3.4   Pharmacokinetics of ghrelin

Mean plasma concentration-time profiles of rat ghrelin after intravenous administration of 100 µg in male rats (Figure 15-5) demonstrate that plasma ghrelin levels by C-RIA increased about 1.5-fold at 1 minute after ghrelin injection. Levels assessed by C-RIA were sustained better than those measured by N-RIA. The half-life of rat ghrelin after intravenous administration was 8 minutes and 24 minutes by N-RIA and C-RIA, respectively.

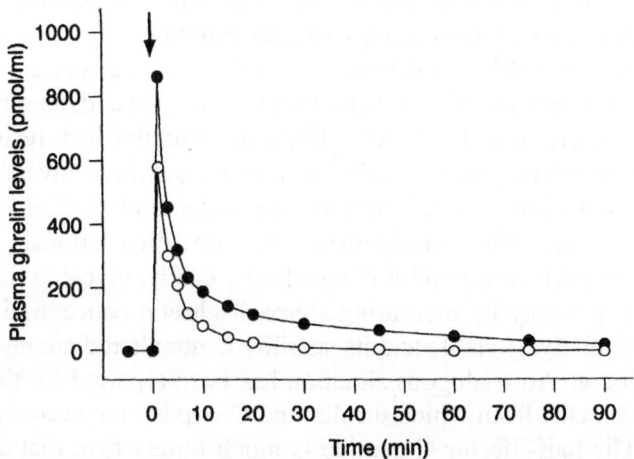

Figure 15-5. Plasma concentration after intravenous administration of rat ghrelin to anesthetized rats. The arrow indicates a bolus injection of ghrelin (100 µg/body) or saline. Mead plasma concentration-time profiles were assessed by C-RIA (●) and N-RIA (O). Each point represents the average of 3 individual samples.

4. DISCUSSION

To distinguish the active form of ghrelin, we established two ghrelin-specific RIAs; N-RIA recognizes the N-terminal, octanoyl-modified portion of the peptide, while C-RIA recognizes the C-terminal portion. Thus, the value determined by N-RIA specifically measures active ghrelin, while the value determined by C-RIA gives the total ghrelin concentration, including both active and des-acyl ghrelin. Both forms of the ghrelin peptide, acylated ghrelin and des-acyl ghrelin, exist in human and rat plasma (19,21). The proportion of active ghrelin in plasma was approximately 2-5% in rodents (22) and 10% in human (Hosoda, unpublished results), implying that the inactive, des-acyl ghrelin is present in the bloodstream at much higher levels than active ghrelin. We therefore examined the pharmacokinetics of rat ghrelin by both N- and C-RIA after intravenous administration. Both plasma ghrelin forms had short half-lives; signals, however, disappeared earlier in N-RIA than in C-RIA, suggesting that there may be differential rates of metabolic turnover for acylated and des-acyl ghrelin. As deletion of the octanoyl modification of ghrelin occurred during the storage of culture medium (18), it is likely that octanoylated ghrelin degraded into des-acylated ghrelin in whole blood and plasma (data not shown). Thus, "active" ghrelin can be rapidly hydrolyzed in circulating blood into "non-active" des-acyl ghrelin, probably through the cleavage of ester bonding.

Plasma ghrelin values obtained by N-RIA demonstrated that the magnitude of plasma ghrelin changes more rapidly and dynamically than those values determined by C-RIA (Hosoda, unpublished results). The measurement of active ghrelin facilitates a more sensitive investigation of small changes in ghrelin levels which may have profound physiological effects. Moreover, the mechanism of post-translational octanoyl modification, important in ghrelin biosynthesis, can be revealed through the use of N-RIA, possibly by measuring stomach ghrelin concentrations. Des-acyl ghrelin is relatively stable and its stability is not altered among different storage conditions. An analogous situation has been reported for the activity of pancreatic b cells, from which insulin and C-peptide are secreted in a 1:1 molar ratio. The half-life for C-peptide is much longer than that of insulin, however, leaving more C-peptide available in the circulation for quantification (23,24). Measuring C-peptide levels provides an assessment of beta-cell secretory activity. Similarly, des-acyl ghrelin levels may serve as an indicator of ghrelin secretory function (14).

Sample stability must be considered for all quantitative assays as previous studies have demonstrated the influence of storage conditions on the analysis of endocrine substances (25-29). Similar to previous findings reporting ghrelin concentrations by C-RIA (30), we demonstrated that, in whole blood and plasma, acylated ghrelin is unstable. The octanoyl modification may be degraded by at least two mechanisms involving either chemical spontaneous hydrolysis or enzymatic cleavage of ester bonding. To obtain accurate data on ghrelin concentrations, blood samples should be centrifuged as soon as possible, at latest within 30 minutes after collection. Acidification is a simple, reliable procedure protecting against the degradation of acylated modification, giving highly improved stability at pH 3-4. This study thus recommends a standard procedure for the collection of blood samples: EDTA-aprotinin treatment and storage of acidified plasma under cooled condition is recommended prior to assessment, to reduce acylated ghrelin degradation to a minimum.

In the sample pretreatment step for RIA systems, extraction of peptide from blood plasma is required (12,15,19). The extraction procedure takes a great deal of time and money. We have developed a new assay system, enzyme immunoassay (EIA), which makes possible the direct measurement of plasma ghrelin levels without the extraction procedure. This EIA system might make it increasingly of interest to explore the role of ghrelin in the clinical setting as a potential diagnostic tool.

REFERENCES

1. Kojima M, Hosoda H, Date Y, Nakazato M, Matsuo H, Kangawa K. Ghrelin is a growth-hormone-releasing acylated peptide from stomach. Nature. 1999; 402:656-60.
2. Kojima M, Hosoda H, Matsuo H, Kangawa K. Ghrelin: discovery of the natural endogenous ligand for the growth hormone secretagogue receptor. Trends Endocrinol Metab. 2001; 12:118-22.
3. Howard AD, Feighner SD, Cully DF, et al. A receptor in pituitary and hypothalamus that functions in growth hormone release. Science. 1996; 273:974-7.
4. McKee KK, Palyha OC, Feighner SD, et al. Molecular analysis of rat pituitary and hypothalamic growth hormone secretagogue receptors. Mol Endocrinol. 1997; 11:415-23.
5. Smith RG, Van der Ploeg LH, Howard AD, et al. Peptidomimetic regulation of growth hormone secretion. Endocr Rev. 1997; 18:621-45.
6. Takaya K, Ariyasu H, Kanamoto N, et al. Ghrelin strongly stimulates growth hormone release in humans. J Clin Endocrinol Metab. 2000; 85:4908-11.
7. Seoane LM, Tovar S, Baldelli R, et al. Ghrelin elicits a marked stimulatory effect on GH secretion in freely-moving rats. Eur J Endocrinol. 2000; 143:R7-9.
8. Peino R, Baldelli R, Rodriguez-Garcia J, et al. Ghrelin-induced growth hormone secretion in humans. Eur J Endocrinol. 2000; 143:R11-4.
9. Tschop M, Smiley DL, Heiman ML. Ghrelin induces adiposity in rodents. Nature. 2000; 407:908-13.
10. Nakazato M, Murakami N, Date Y, et al. A role for ghrelin in the central regulation of feeding. Nature. 2001; 409:194-8.
11. Inui A. Ghrelin: an orexigenic and somatotrophic signal from the stomach. Nat Rev Neurosci. 2001; 2:551-60.
12. Ariyasu H, Takaya K, Tagami T, et al. Stomach is a major source of circulating ghrelin, and feeding state determines plasma ghrelin-like immunoreactivity levels in humans. J Clin Endocrinol Metab. 2001; 86:4753-8.
13. Tschop M, Weyer C, Tataranni PA, Devanarayan V, Ravussin E, Heiman ML. Circulating ghrelin levels are decreased in human obesity. Diabetes. 2001; 50:707-9.
14. Cummings DE, Purnell JQ, Frayo RS, Schmidova K, Wisse BE, Weigle DS. A preprandial rise in plasma ghrelin levels suggests a role in meal initiation in humans. Diabetes. 2001; 50:1714-9.
15. Shiiya T, Nakazato M, Mizuta M, et al. Plasma ghrelin levels in lean and obese humans and the effect of glucose on ghrelin secretion. J Clin Endocrinol Metab. 2002; 87:240-4.
16. Hansen TK, Dall R, Hosoda H, et al. Weight loss increases circulating levels of ghrelin in human obesity. Clin Endocrinol (Oxf). 2002; 56:203-6.
17. Nagaya N, Uematsu M, Kojima M, et al. Elevated circulating level of ghrelin in cachexia associated with chronic heart failure: relationships between ghrelin and anabolic/catabolic factors. Circulation. 2001; 104:2034-8.
18. Kanamoto N, Akamizu T, Hosoda H, et al. Substantial production of ghrelin by a human medullary thyroid carcinoma cell line. J Clin Endocrinol Metab. 2001; 86:4984-90.
19. Hosoda H, Kojima M, Matsuo H, Kangawa K. Ghrelin and des-acyl ghrelin: two major forms of rat ghrelin peptide in gastrointestinal tissue. Biochem Biophys Res Commun. 2000; 279:909-13.
20. Date Y, Kojima M, Hosoda H, et al. Ghrelin, a novel growth hormone-releasing acylated peptide, is synthesized in a distinct endocrine cell type in the gastrointestinal tracts of rats and humans. Endocrinology. 2000; 141:4255-61.

21. Hosoda H, Kojima M, Mizushima T, Shimizu S, Kangawa K. Structural divergence of human ghrelin. Identification of multiple ghrelin-derived molecules produced by post-translational processing. J Biol Chem. 2003; 278:64-70.
22. Ariyasu H, Takaya K, Hosoda H, et al. Delayed short-term secretory regulation of ghrelin in obese animals: evidenced by a specific RIA for the active form of ghrelin. Endocrinology. 2002;143:3341-50.
23. Myrick JE, Gunter EW, Maggio VL, Miller DT, Hannon WH. An improved radioimmunoassay of C-peptide and its application in a multiyear study. Clin Chem. 1989; 35:37-42.
24. Horwitz DL, Starr JI, Mako ME, Blackard WG, Rubenstein AH. Proinsulin, insulin, and C-peptide concentrations in human portal and peripheral blood. J Clin Invest. 1975; 55:1278-83.
25. Nelesen RA, Dimsdale JE, Ziegler MG. Plasma atrial natriuretic peptide is unstable under most storage conditions. Circulation. 1992; 86:463-6.
26. Flower L, Ahuja RH, Humphries SE, Mohamed-Ali V. Effects of sample handling on the stability of interleukin 6, tumour necrosis factor-alpha and leptin. Cytokine. 2000; 12:1712-6.
27. Boomsma F, Alberts G, van Eijk L, Man in 't Veld AJ, Schalekamp MA. Optimal collection and storage conditions for catecholamine measurements in human plasma and urine. Clin Chem. 1993; 39:2503-8.
28. Miki K, Sudo A. Effect of urine pH, storage time, and temperature on stability of catecholamines, cortisol, and creatinine. Clin Chem. 1998; 44(8 Pt 1):1759-62.
29. Evans MJ, Livesey JH, Ellis MJ, Yandle TG. Effect of anticoagulants and storage temperatures on stability of plasma and serum hormones. Clin Biochem. 2001; 34:107-12.
30. Groschl M, Wagner R, Dotsch J, Rascher W, Rauh M. Preanalytical influences on the measurement of ghrelin. Clin Chem. 2002; 48:1114-6.

Chapter 16

GHRELIN: IMPLICATIONS IN PEDIATRIC ENDOCRINOLOGY

Simonetta Bellone, Anna Rapa, Fabio Broglio[1] & Gianni Bona
Unit of Pediatrics, Department of Medical Sciences, University of Piemonte Orientale, Novara; [1]Division of Endocrinology, Department of Internal Medicine, University of Turin, Turin, Italy.

Abstract: Ghrelin, a 28 amino-acid acylated peptide predominantly produced by the stomach, displays strong growth hormone (GH)-releasing activity mediated by the hypothalamus-pituitary GH Secretagogue (GHS) receptors which had been shown to be specific for a family of synthetic, orally active GHS. Ghrelin and GHS show other endocrine and nonendocrine actions including orexigenic effects and influence on gastro-entero-pancreatic functions. Ghrelin manages the neuroendocrine and metabolic response to starvation. The study of ghrelin secretion as function of age and gender as well as the study of the endocrine and nonendocrine effects of ghrelin and its analogues in physiological and pathological conditions will likely provide critical information about the role of ghrelin and the potential perspectives of its analogues in clinical practice. This point is of particular interest in the field of pediatric endocrinology and metabolism because the ghrelin story started focusing on GH deficiency and is now extending to aspects that once again are of major relevance such as obesity and eating disorders, regulation of the hypothalamus-pituitary-adrenal and gonadal axis. GHS analogues acting as agonists or antagonists on appetite could represent new drug intervention in eating disorders. GHS could therefore represent a reliable provocative test for the diagnosis of GH deficiency but as orally active growth-promoting agents they are not comparable with rhGH in terms of efficacy.

Key words: GH, IGF-I, genetic, GH deficiency, obesity

1.   INTRODUCTION

Ghrelin, a 28 amino-acid peptide predominantly produced by the stomach, displays strong growth hormone (GH)-releasing activity mediated by the activation of the GH Secretagogues (GHS) receptors (GHS-R) type 1a which is specific for synthetic, peptidyl and non peptidyl GHS (1-4). GHS were invented more than 20 years ago and their strong GH-releasing activity even after oral administration suggested potential clinical usefulness for diagnosis and treatment of GH deficiency (GHD) in childhood as well as for treatment of other conditions of GH insufficiency including aging (2,4). Ghrelin and synthetic GHS act via receptors concentrated in the hypothalamus-pituitary unit but are also distributed in other central and peripheral tissues (2,5). While hypothalamus-pituitary receptors explain the stimulatory effect of GHS on GH and also on prolactin and adrenocorticotropic hormone secretion as well as the inhibitory influence on gonadal axis, other central and peripheral specific binding sites explain other activities (1,4,6-9). Among these, the orexant activity coupled with control of energy expenditure is receiving particular interest because of the potential therapeutic implications for ghrelin analogues acting as agonists and antagonists.

The study of ghrelin secretion as function of age and gender as well as the study of the endocrine and nonendocrine effects of ghrelin and its analogues in physiological and pathological conditions will likely provide critical information about the role of ghrelin and the potential perspectives of its analogues in clinical practice. This point is of particular interest in the field of pediatric endocrinology and metabolism because the ghrelin story started focusing on GHD and is now extending to areas that are of major relevance today such as obesity and eating disorders, and regulation of the hypothalamus-pituitary-adrenal and -gonadal axis.

Aim of this chapter is to consider the knowledge accumulated so far about ghrelin secretion and actions in childhood and meantime to review the potential clinical pediatric implications of ghrelin analogues.

2.   GHRELIN SECRETION: FROM NEWBORNS TO ADULTHOOD

Ghrelin secretion, mostly represented in its acylated form, occurs in pulsatile manner. It is noteworthy that there is no strict correlation between ghrelin and GH levels while ghrelin pulses are correlated with food intake episodes and sleep cycles in rats (10). Particularly, in humans it has been shown that peaks in ghrelin levels anticipate food intake suggesting the latter is triggered by ghrelin discharge (11).

In agreement with the major influence of nutrition on ghrelin secretion, circulating ghrelin levels are increased in anorexia and cachexia but reduced in obesity and overfeeding (12). In all these conditions ghrelin secretion is normalized by recovery of ideal body weight (13). These changes are opposite to those of leptin suggesting that both ghrelin and leptin are hormones signaling the metabolic balance and managing the neuro-endocrine and metabolic response to starvation (12,14).

Circulating ghrelin levels mostly reflect gastric secretion; in fact, they are reduced by 70% after gastrectomy (15) though this reduction has been reported transient (16).

In humans, ghrelin levels are increased by fasting and energy restriction but decreased by food intake and overfeeding (12). It had been reported that ghrelin secretion is not inhibited by simple gastric distension in animals but, more recently, it has been demonstrated that gastric bypass strongly inhibits it (17). Both oral and intravenous glucose loads inhibit ghrelin secretion in humans as well as in animals; on the other hand, intravenous free fatty acid load as well as arginine load do not affect circulating ghrelin levels (18).

In agreement with the negative association between ghrelin secretion and body mass, clear negative association between ghrelin and insulin secretion has been found in humans as well as in animals (11,19,20) suggesting an inhibitory influence of insulin on ghrelin secretion (21,22). Indeed, during an euglycemic clamp the steady state increase in insulin levels is associated with clear reduction in circulating ghrelin levels (18). The exact mechanisms by which insulin and glucose variations regulate ghrelin secretion is still unknown though it has been demonstrated that insulin directly modulates ghrelin expression at the gastric level (23).

The most remarkable inhibitory input on ghrelin secretion is represented by the activation of somatostatin receptors done by somatostatin as well as by its natural analogue cortistatin that, meantime, reduce beta-cell secretion (24). Somatostatin receptor subtypes are present in the gastric mucosa and mostly sst2 and sst5 receptors, which would mediate the inhibitory influence on ghrelin secretion (25,26).

In all, ghrelin secretion is under major influence, mostly inhibitory, from the endocrine pancreas. However, despite the strict negative association between insulin and ghrelin, obese patients with Prader-Willi syndrome show absolute or at least relative ghrelin hypersecretion (27,28). This peculiar exception to the rule that ghrelin secretion is negatively associated to body mass index as well as to insulin suggests that overweight in this syndrome could be the result of exaggerated food intake triggered by the orexigenic effect of ghrelin hypersecretion (28).

The influence of age and gender on ghrelin secretion has not been definitively clarified. At present ghrelin secretion seems independent of gender in both adults and children, and even in newborns (20). This assumption, however, is based on studies in which ghrelin levels have only been measured in the morning; appropriate studies addressing more prolonged evaluation of ghrelin secretion are therefore needed to definitely rule out any impact of gender on the secretion of this gastric hormone.

Data regarding ghrelin secretion as function of aging from fetal and neonatal life to childhood and adolescence up to adulthood is scanty. In the following paragraphs we will address this point basing on our recent studies.

2.1   Newborns

The control of growth in newborns is different from that in later life and involves genetic, nutritional, hormonal and environmental factors. In the fetus as well as in the newborn, insulin, insulin-like growth factor-I (IGF-I) and nutrient availability are the main determinants of growth while GH secretion does not play a major role (29,30). GH levels are high during the fetal and neonatal periods and this hypersecretory state likely reflects blunted negative feedback action of IGF-I due to peripheral GH insensitivity (31). The role, if any, of the ghrelin system in driving GH secretion during fetal life and at birth is still unclear. Recently it has been reported that ghrelin levels in newborns show wide variability and is surprisingly negatively associated to GH and positively to leptin levels (32).

We studied ghrelin levels in 93 full-term newborns appropriate for gestational age, their mothers immediately after delivery, and 19 lean healthy adults as comparison. The results demonstrated that newborns show ghrelin levels similar to adults; this age-related independency of ghrelin levels is at variance with the age-related variations in GH levels that are higher at birth.

Peculiarly, ghrelin secretion at birth, at least in full-term appropriate for gestational age newborns, was independent of body weight and length, and IGF-I, GH, insulin and leptin levels. Therefore, these findings seem contrary to possibility that GH hypersecretion at birth is driven by ghrelin whose levels are not associated to GH levels (33).

We also found that ghrelin secretion at birth is independent of body weight and length and other hormonal levels including IGF-I, insulin and leptin secretion (33). This lack of any correlation between ghrelin levels and metabolic or anthropometric parameters suggests that this new gastric hormone does not have at birth the same metabolic activities as in adulthood when clear association with body mass index, nutrient intake, and insulin secretion is clearly apparent (12,19).

Interestingly, ghrelin levels in newborns were higher than those in their mothers indicating feto-placental-derived ghrelin production (33). Ghrelin expression in the rat and human placenta has been demonstrated in the first half of pregnancy with almost undetectable levels at term (34) suggesting that changes in cord ghrelin levels mostly depends on age-related fetal production.

In all, ghrelin secretion in newborns is similar to that in adults in agreement with the assumption that it does not undergo relevant age-related variations. Notably, the ghrelin secretory pattern is not linked with that of GH, whose secretion is maximal at birth. Thus, it seems unlikely that ghrelin plays major role in the physiological control of GH secretion, at least at birth. Moreover, evidence that ghrelin levels in full term newborns with appropriate gestational age do not correlate with either IGF-I or body weight, length, or insulin and leptin levels, bringing into question this gastric hormone's role in fetal body growth.

2.2 Children

It is widely accepted that GH and IGF-I secretion increases at puberty and that normal growth in childhood and adolescence depends on the appropriate pulsatile GH secretion that, in turn, generates IGF-I secretion (35). The age-related variations of GH secretion throughout childhood mostly depend on the hypothalamic control of somatotroph function, particularly on the activity of growth hormone releasing hormone (GHRH)-secreting neurons and their interplay with somatostatin (35).

Following the discovery of synthetic GHS, of its specific receptors and finally of the endogeneous ligand ghrelin, it has been hypothesized that, besides GHRH, somatostatin, neurotransmitters, peripheral hormones and metabolic fuels, the ghrelin system would play a major role in the age-related variations of GH secretion across the lifespan.

We first measured morning ghrelin levels after overnight fasting in 41 healthy lean children and 39 obese children either in prepubertal or pubertal age comparing them with those recorded in a group of young adults of both sexes. Morning ghrelin levels after overnight fasting were similar between males and females as well as between adults and children. Ghrelin secretion was also independent of puberty (20). In agreement with previous reports in adults (19,12), in obese children ghrelin levels were found reduced and negatively correlated to overweight and insulin secretion.

In this study we showed that ghrelin levels in adults and in children are gender-independent, though GH secretion is enhanced in females, at least in adulthood (35). Differently from GH secretion that increases at puberty and decreases thereafter, ghrelin levels in children do not increase at puberty and are similar to those in adults (20). Indeed in the present study ghrelin levels were even negatively associated to IGF-I levels, the best marker of GH status that undergoes a well-known increase at puberty (35).

Besides stimulating GH, ghrelin exerts an orexigenic effect and manages the endocrine and metabolic response to variations in the nutritional status

(11,12,14). Evidence of reduced ghrelin levels in obese children and adults is contrary to the hypothesis that obesity could reflect ghrelin-stimulated overfeeding (12,14), the notable exception being Prader-Willi syndrome that is connoted by peculiar absolute or relative elevation of ghrelin secretion (27,28).

· In all, ghrelin levels in childhood are similar to those in adulthood and do not depend on gender but are negatively associated to overweight that, therefore, is confirmed as one of the most remarkable variables influencing its secretion.

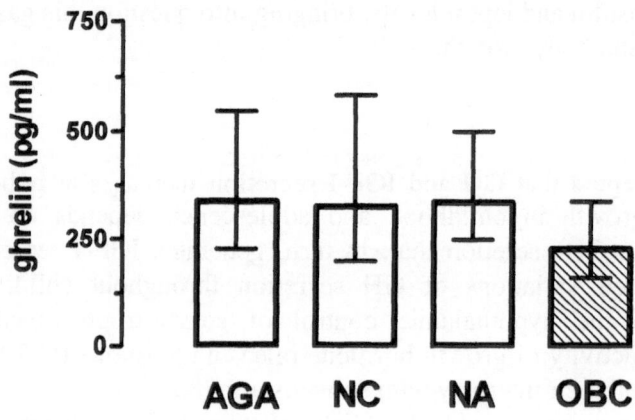

Figure 16-1. Ghrelin levels in neonates (AGA), lean children (NC), lean adults (NA) and obese children (OBC)

3.   GENETIC ANALYSIS

Considerable interest has been raised by the possibility that ghrelin could be implicated in the etiology of obesity and that mutations in the ghrelin gene could be associated with obesity in humans.

Ghrelin was purified from rat stomach, and the gene was subsequently cloned in rats and humans (1). The human ghrelin gene is located on chromosome 3, at locus 3p25-26, and is made up of 4 exons and 3 introns (GenBank= AF296558) The heteronuclear RNA of the gene transcript is processed by alternative splicing to yield two different mature mRNAs: one that produces the ghrelin precursor of 117 amino acids, and another one that produces a different form, with similar biological activities (des-Gln 14-ghrelin) (16). Moreover, before being secreted, the ghrelin molecule undergoes an enzymatic process in the cytoplasm, a *n*-octanoyl addition at serine 3, that is essential for the biological activity and for the *in vitro* binding to the subtype-1a of the GHS-R.

Recently, three polymorphisms in the ghrelin gene have been reported from different studies (36,37*)*. The first is a single base substitution G152A that leads to a replacement of arginine (Arg) by glutamine (Gln) at codon 28, that is the last codon of the mature ghrelin product, whereas the second and the third variants are in the prepro-ghrelin sequence and do not affect the mature product (C214A exchanges leucine (Leu) by methionine (Met) at codon 72, and A269T exchanges Gln by Leu at codon 90). Interestingly, (36,38) the variation 72Met has been initially reported to be associated with an earlier age of self-reported onset of obesity, while following studies (39) proposed that the Leu72Met status could be protective against fat accumulation and associated metabolic comorbidities, although the mechanisms for these associations are unknown. Instead, neither the Arg51Gln nor the Gln90Leu variants seem to influence weight regulation, even if the ghrelin Arg51Gln was associated with lower plasma ghrelin levels than the Arg51Arg status.

Other studies are therefore needed to test whether genetic variants in the ghrelin gene could play a role in predisposition to early obesity onset or in modulating some aspects of the obese phenotype during childhood and to test whether polymorphisms could influence plasma ghrelin levels.

4. CLINICAL IMPLICATIONS OF GH SECRETAGOGUES AND GHRELIN IN CHILDHOOD

Potential clinical targets for ghrelin and its synthetic analogues in pediatric endocrinology would theoretically include growth disorders due to GH insufficiency, obesity and eating disorders. In fact, ghrelin analogues acting as agonists or antagonists and able to enhance or reduce appetite and food intake would have obvious usefulness in anorectic and obese patients, respectively. At present this possibility is not supported by any evidence. On the other hand, the only clinical implications for ghrelin analogues, i.e. synthetic GHS, would attain to growth disorders.

4.1 Diagnostic use

The stimulatory effect of ghrelin on GH secretion in humans has been demonstrated even in GHD patients. In fact, in adults with childhood-onset GHD, ghrelin significantly stimulates GH levels although to an extent dramatically lower than in normal subjects (40). Noteworthy, in GHD, ghrelin stimulates GH secretion more than in an insulin-tolerance test and even than in the GHRH+arginine test (40).

GHS as well as ghrelin represent a potent and reproducible stimulus of GH secretion in children as well as in adults. The GH-releasing effect of GHS does not depend on sex but undergoes marked age-related variations. It is present at birth and persists in prepubertal children, then it clearly increases at puberty, persisting into adulthood and decreasing with aging (4,41). The age-related effect of GHS is quantitatively and even qualitatively different from that of GHRH whose effect seems maximal at birth and then progressively decreases throughout aging.

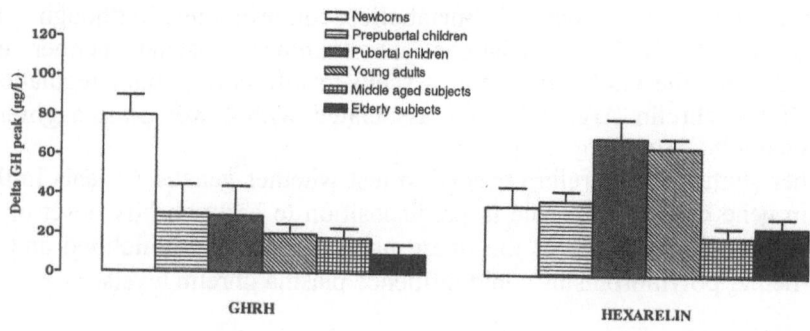

Figure 16-2. GH response to GHRH (1.0 µg/kg) and Hexarelin (1.0 µg/kg) test in different ages of life.

The mechanisms underlying the age-related variations in the GH-releasing activity of GHS differ age by age. The enhanced GH-releasing effect of GHS at puberty reflects the positive influence of estrogens. In fact the GH response to GHS is more marked in pubertal girls than in boys is positively related to estradiol levels; moreover the GH rise in prepubertal children is enhanced by testosterone as well as by ethynyl-estradiol but not by oxandrolone pretreatment (4,42).

The diagnosis of GHD is established for the first step by the normal response to provocative tests of GH secretion. Classical tests are connoted by a high variability in GH response while GHS shows a limited intraindividual variability. Moreover, when combined with GHRH, GHS represents one of the most potent and reliable tests to evaluate the pituitary GH releasable pool for the diagnosis of GHD (41,43,44). It is widely accepted that the diagnosis of GHD in childhood is not simply assessed by the GH response to either classical or maximal provocative tests (45). A considerable number of short children with normal GH response to provocative tests shows insufficient daily GH secretion reflecting neuro-secretory dysfunction and benefit from GH replacement (45). Though normal GH response to stimuli does not rule out GH insufficiency in childhood, potent and reproducible provocative tests of GH secretion such as natural or synthetic GHS either alone or combined with GHRH can provide definite information about the maximal secretory capacity of somatotroph cells in short children suspected for GHD (41,45).

4.2 Therapeutic use

Based on their strong and reproducible GH-releasing effects even after oral administration, GHS could theoretically have diagnostic and therapeutic usefulness. The possibility that GH-releasing substances, particularly if orally active, could represent a therapeutic approach alternative to recombinant human GH (rhGH) in GHD patients received considerable attention.

Clearly, GH-releasing substances have no place as alternatives to rhGH for treatment of severe GHD in patients with panhypopituitarism due to massive destruction of the pituitary gland. On the other hand, isolated GHD often reflects hypothalamic pathogenesis, as shown by clear GH responses to GHRH in many dwarf patients (45). It was hypothesized that patients with isolated GHD could benefit from treatment with GHRH and preferably with orally active GHS, having the possible advantage of restoring endogenous GH pulsatility, and therefore being a more "physiological" approach.

The potential usefulness of GHS for treatment of short stature with isolated GHD was suggested by some open studies reporting increases in height velocity in short children with idiopathic short stature or GHD after chronic treatment with intranasal or subcutaneous GHS (46-49). When tested in double blind, placebo-controlled trials in short children with GHD, the efficacy of the most promising orally active GHS, MK-0677, was found not comparable with that of rhGH (50). Treatment with MK-0677 transiently increased height velocity (approximately 3 cm/year) in dose-independent manner in children with partial GHD but not in children with severe GHD, who had no benefit at all by the treatment (50). Thus, it is hard to believe that GHS can replace rhGH for treatment of GHD in childhood. This evidence was also contrary to the hypothesis that isolated GHD could reflect defects in the activity of the endogenous GHS-like ligand, i.e., ghrelin. As a general comment on the failure of GHS as alternative to rhGH for treatment of GHD, it can be said that the clear dependence of GHS activity on the full integrity of the hypothalamus-pituitary function could predict their failure in pathophysiologic conditions that, by definition, include alteration in the hypothalamus-pituitary unit.

REFERENCES

1. Kojima M, Hosoda H, Matsuo H, Kangawa K. Ghrelin: discovery of the natural endogenous ligand for the growth hormone secretagogue receptor. Trends Endocrinol Metab. 2001; 12:118-22.
2. Dieguez C, Casanueva FF. Ghrelin: a step forward in the understanding of somatotroph cell function and growth regulation. Eur J Endocrinol. 2000; 142:413-7.
3. Kojima M, Hosoda H, Date Y, et al. Ghrelin is a growth-hormone-releasing acylated peptide from stomach. Nature. 1999; 402:656-60.
4. Ghigo E, Arvat E, Giordano R, et al. Biologic activities of growth hormone secretagogues in humans. Endocrine. 2001; 14:87-93.
5. Papotti M, Cassoni P, Volante M, et al. Ghrelin-producing endocrine tumors of the stomach and intestine. J Clin Endocrinol Metab. 2001; 86:5052-9.
6. Arvat, E, Maccario M, Di Vito L, et al. Endocrine activities of ghrelin, a natural growth hormone secretagogue (GHS), in humans: comparison and interactions with hexarelin, a nonnatural peptidyl GHS, and GH-releasing hormone. J Clin Endocrinol Metab. 2001; 86:1169-74.
7. Takaya K, Ariyasu H, Kanamoto N, et al. Ghrelin strongly stimulates growth hormone release in humans. J Clin Endocrinol Metab. 2000; 85:4908-11.
8. Tena-Sempere M, Barreiro ML, Gonzalez LC, et al. Novel expression and functional role of ghrelin in rat testis. Endocrinology. 2002; 143:717-25.
9. Furuta M, Funabashi T, Kimura F. Intracerebroventricular administration of ghrelin rapidly suppresses pulsatile luteinizing hormone secretion in ovariectomized rats. Biochem Biophys Res Commun. 2001; 288:780-5.
10. Tolle V, Bassant MH, Zizzari P, et al. Ultradian rhythmicity of ghrelin secretion in relation with GH, feeding behavior, and sleep-wake patterns in rats. Endocrinology. 2002; 143:1353-61.
11. Cummings DE, Purnell JQ, Frayo RS, Schmidova K, Wisse BE, Weigle DS. A preprandial rise in plasma ghrelin levels suggests a role in meal initiation in humans. Diabetes. 2001; 50:1714-9.
12. Inui A. Ghrelin: an orexigenic and somatotrophic signal from the stomach. Nat Rev Neurosci. 2001; 2:551-60.
13. Otto B, Cuntz U, Fruehauf E, et al. Weight gain decreases elevated plasma ghrelin concentrations of patients with anorexia nervosa. Eur J Endocrinol. 2001; 145:669-73.
14. Horvath TL, Diano S, Sotonyi P, et al. Ghrelin and the regulation of energy homeostasis: A hypothalamic perspective. Endocrinology. 2001;142:4163-9.
15. Ariyasu H, Takaya K, Tagami T, et al. Stomach is a major source of circulating ghrelin, and feeding state determines plasma ghrelin-like immunoreactivity levels in humans. J Clin Endocrinol Metab. 2001; 86:4753-8.
16. Hosoda H, Kojima M, Mizushima T, Shimizu S, Kangawa K. Structural divergence of human ghrelin. Identification of multiple ghrelin-derived molecules produced by post-translational processing. J Biol Chem. 2003; 278:64-70.
17. Cummings DE, Weigle DS, Frayo RS, et al. Plasma ghrelin levels after diet-induced weight loss or gastric bypass surgery. N Engl J Med. 2002; 346:1623-30.
18. Mohlig M, Spranger J, Otto B, Ristow M, Tschop M, Pfeiffer AF. Euglycemic hyperinsulinemia, but not lipid infusion, decreases circulating ghrelin levels in humans. J Endocrinol Invest. 2002; 25:RC36-8.

19. Tschop M, Weyer C, Tataranni PA, Devanarayan V, Ravussin E, Heiman ML. Circulating ghrelin levels are decreased in human obesity. Diabetes. 2001; 50:707-9.

20. Bellone S, Rapa A, Vivenza D, et al. Circulating ghrelin levels as function of gender, pubertal status and adiposity in childhood. J Endocrinol Invest. 2002; 25:RC13-5.

21. Saad MF, Bernaba B, Hwu CM, et al. Insulin regulates plasma ghrelin concentration. J Clin Endocrinol Metab. 2002; 87:3997-4000.

22. Lucidi P, Murdolo G, Di Loreto C, et al. Ghrelin is not necessary for adequate hormonal counterregulation of insulin-induced hypoglycemia. Diabetes. 2002; 51:2911-4.

23. Lee HM, Wang G, Englander EW, Kojima M, Greeley GH Jr. Ghrelin, a new gastrointestinal endocrine peptide that stimulates insulin secretion: enteric distribution, ontogeny, influence of endocrine, and dietary manipulations. Endocrinology. 2002; 143:185-90.

24. Broglio F, van Koetsveld P, Benso A, et al. Ghrelin secretion is inhibited by either somatostatin or cortistatin in humans. J Clin Endocrinol Metab. 2002; 87:4829-32.

25. Patel YC. Somatostatin and its receptor family. Front Neuroendocrinol. 1999; 20:157-98.

26. Norrelund H, Hansen TK, Orskov H, et al. Ghrelin immunoreactivity in human plasma is suppressed by somatostatin. Clin Endocrinol (Oxf). 2002; 57:539-46.

27. Haqq AM, Farooqi IS, O'Rahilly S, et al. Serum ghrelin levels are inversely correlated with body mass index, age, and insulin concentrations in normal children and are markedly increased in Prader-Willi syndrome. J Clin Endocrinol Metab. 2003; 88:174-8.

28. DelParigi A, Tschop M, Heiman ML, et al. High circulating ghrelin: a potential cause for hyperphagia and obesity in prader-willi syndrome. J Clin Endocrinol Metab. 2002; 87:5461-4.

29. Ogilvy-Stuart AL, Hands SJ, Adcock CJ, et al. Insulin, insulin-like growth factor I (IGF-I), IGF-binding protein-1, growth hormone, and feeding in the newborn. J Clin Endocrinol Metab. 1998; 83:3550-7.

30. Wollmann HA. Growth hormone and growth factors during perinatal life. Horm Res. 2000; 53:50-4.

31. Gluckman PD. The endocrine regulation of fetal growth in late gestation. The role of insulin-like growth factors. J. Clin. Endocrinol. Metab. 1995; 80:1047–50.

32. Chanoine JP, Yeung LP, Wong AC, Birmingham CL. Immunoreactive ghrelin in human cord blood: relation to anthropometry, leptin, and growth hormone. J Pediatr Gastroenterol Nutr. 2002; 35:282-6.

33. Bellone S, Rapa A, Vivenza D, et al. Circulating ghrelin levels in newborns are not associated to gender, body weight and hormonal parameters but depend on the type of delivery. J Endocrinol Invest, 2003; 26:RC9-11.

34. Gualillo O, Caminos J, Blanco M, et al. Ghrelin, a novel placental-derived hormone. Endocrinology. 2001; 142:788-94.

35. Ghigo E., Arvat E., Gianotti L., Maccario M., Camanni F. The regulation of growth hormone secretion. In: Jenkins RC, Ross RJM, eds. The endocrine response to acute illness. Front Horm Res. Basel: Karger. 1999; 24:152-75.

36. Ukkola O, Ravussin E, Jacobson P, et al. Muations in the preproghrelin/ghrelin gene associated with obesity in humans. J Clin Endocrinol Metab. 2001; 86:3996-9.

37. Hinney A, Hoch A, Geller F, et al. Ghrelin gene: identification of missense variants and a frameshift mutation in extremely obese children and adolescents and healthy normal weight students. J Clin Endocrinol Metab. 2002; 87:2716-9.

38. Korbonits M,Gueorguiev M, O'Grady E, et al. A variation in the ghrelin gene increases weight and decreases insulin in tall, obese children. J Clin Endocrinol Metab. 2002; 87:4005-8.

39. Ukkola O, Ravussin E, Jacobson P, et al. Role of ghrelin polymorphisms in obesity based on three different studies. Obes Res. 2002; 10:782-91.

40. Aimaretti G, Baffoni C, Broglio F, et al. Endocrine responses to ghrelin in adult patients with isolated childhood-onset growth hormone deficiency. Clin Endocrinol (Oxf). 2002; 56:765-71.

41. Ghigo E, Arvat E, Aimaretti G, et al. Diagnostic and therapeutic uses of growth hormone-releasing substances in adult and elderly subjects. Baillière's Clin Endocrinol Metab. 1998; 12:341-58.

42. Loche S, Colao A, Cappa M, et al. The growth hormone response to hexarelin in children: reproducibility and effect of sex steroids. J Clin Endocrinol Metab. 1997; 82:861-4.

43. Popovic V, Leal A, Micic D, et al. GH-releasing hormone and GH-releasing peptide-6 for diagnostic testing in GH-deficient adults. Lancet 2000; 356:1137-42.

44. Laron Z. Intranasally and orally active GH secretagogues are useful clinical tools: so why are they not on the market? J Endocrinol Invest. 2003; 26:91-2.

45. Shalet SM, Toogood A, Rahim A, Brennan BM. The diagnosis of growth hormone deficiency in children and adults. Endocr Rev. 1998; 19:203-23.

46. Laron Z, Frenkel J, Deghenghi R, et al. Intranasal administration of the GHRP hexarelin accelerates growth in short children. Clin Endocrinol (Oxf). 1995; 43:631-5.

47. Klinger B, Silbergeld A, Deghenghi R, et al. Desensitization from long-term intranasal treatment with hexarelin does not interfere with the biological effects of this growth hormone-releasing peptide in short children. Eur J Endocrinol. 1996; 134:716-9.

48. Mericq V, Cassorla F, Garcia H, et al. Growth hormone (GH) responses to GH-releasing peptide and to GH-releasing hormone in GH-deficient children. J Clin Endocrinol Metab. 1995; 80:1681-4.

49. Pihoker C, Badger TM, Reynolds GA, Bowers CY. Treatment effects of intranasal growth hormone releasing peptide-2 in children with short stature. J Endocrinol. 1997; 155:79-86.

50. Yu H, Cassorla F, Tiulpakov A, et al. A double-blind placebo-controlled efficacy trial of an oral growth hormone (GH) secretagogue (MK-0677) in GH deficient (GHD) children. Program of the 70th Annual Meeting of The Endocrine Society, New Orleans, LA, USA, 1998; p 84 (abstract).

Chapter 17

GHRELIN AND ITS ANALOGUES: PERSPECTIVES FOR A STORY OF REVERSE PHARMACOLOGY

Ezio Ghigo

Division of Endocrinology and Metabolism, Department of Internal Medicine, University of Turin, Turin, Italy

The aim of this book was to provide the reader with an updated review of ghrelin based on the contributions from all the scientists who gave rise to this story; their participation in this book is my personal honor and I am sure that there would have been no other way to address such a wide spectrum of topics.

This concluding paper gives my perspectives for this story of ghrelin and synthetic Growth Hormone Secretagogues (GHS): a story of reverse pharmacology.

GHS were born more than 20 years ago as synthetic, nonnatural molecules possessing growth hormone (GH)-releasing activity *in vitro*. Following molecules were also found to be very active *in vivo* in humans and in animals even after oral administration. This gave rise to the dream that GH deficiency could be treated by orally active GHS as an alternative to recombinant human GH. This hypothesis has never been demonstrated. As aging is connoted by GH insufficiency reflecting age-related changes in hypothalamic control of somatotroph function, that would include some derangement in the "endogenous GHS system", it was also hypothesized that treatment with orally active GHS would be useful as anabolic anti-aging intervention in somatopause. This hypothesis has never been definitely demonstrated either, though the results of some prolonged trials with MK-0677 provided some good perspectives. Again, as obesity is connoted by remarkable GH insufficiency, it was hypothesized that treatment with orally active GHS would increase somatotroph function in obesity providing enhanced lypolisis and probably favoring enhanced fat mass loss during energy restriction; the results of some studies did not support this hypothesis.

Besides these potential clinical implications that stimulated the interest of pharmaceutical companies, the GHS story developed from a more basic point of view up to the discovery of the GHS receptor (GHS-R) that in turn allowed ghrelin's discovery as a natural ligand of the receptors mediating the activities of synthetic GHS. In fact, it was soon demonstrated that GHS possess effects other than those on GH, but extend to other endocrine and even central and peripheral nonendocrine actions. Indeed, it is now clear that the GH-releasing effect of ghrelin, a natural GHS-R ligand, is simply one out of a wide spectrum of actions including: a) stimulation of lactotroph and corticotroph secretion probably coupled with modulatory influence on gonadal axis; b) orexigenic effect coupled with control of energy expenditure; c) influence on sleep and behavior; d) influence on gastro-entero-pancreatic functions and secretions; e) influence on the endocrine pancreatic function as well as on glucose metabolism; f) cardiovascular actions; g) modulation of cell proliferation.

Four years after the discovery of ghrelin, a large part of the literature addressing this new field is mostly focusing on ghrelin influence on central regulation of appetite and energy balance. In fact, like leptin, ghrelin appears as a new hormone signaling the metabolic balance and managing the neuroendocrine and metabolic response to energy balance variations.

Obesity and eating disorders are among the leading causes of illness and mortality in the developed world. To better understand the pathophysiological mechanisms that underlie metabolic disorders, increasing attention has been paid to central regulatory factors in energy homeostasis, including food intake and energy expenditure. As we know, the past two decades have provided overwhelming evidence of the critical role that hypothalamic peptidergic systems play in the central regulation of appetite and energy balance. The discovery of ghrelin and its influence on appetite, fuel utilization, body weight and body composition adds yet another component to the complexity of the central regulation of energy balance; this implication had been anticipated by the pioneer studies of Cyril Bowers who discovered GHS as GH-releasing but also as orexigenic molecules.

Independently of the wide spectrum of pathophysiological aspects in neuroscience, endocrinology, metabolism and internal medicine, whose understanding will be advanced by studies on ghrelin secretion and action, it is clear that further development of the GHS/ghrelin story is now going to be mainly favored by the pharmaceutical dream that orally active, synthetic

GHS analogues acting as agonists or antagonists on appetite and food intake are prospects as drug interventions in eating disorders.

To emphasize the impact of ghrelin on appetite it has recently been shown that even in humans its acute administration enhances appetite and food consumption; this finding is relevant because the majority of peptides believed to have a major influence on food intake generally failed to modify food consumption in humans. Taking into account that synthetic, orally active, peptidyl or non peptidyl GHS acting as agonists or antagonists of ghrelin have already been synthesized and that some of them show a specific action on appetite only, it seems extremely attractive to verify whether their chronic administration is able to modify food consumption, so that body weight in the end increases or decreases. It is easy to foresee that the next few years will be dedicated to clarifying whether this fascinating dream is more realistic than that aiming to generate orally active anabolic agents for anti-aging interventions. Faced with this perspective of basic and clinical research, should we be optimistic or pessimistic? Research needs an optimistic approach and is not simply followed by prompt pharmaceutical applications; there are some findings that would suggest caution. The finding that circulating leptin levels, an anorexant agent, were positively associated to body mass seemed unexpected, like the evidence that ghrelin levels, an orexigenic agent, are reduced in obese and elevated in anorectic patients, i.e., negatively associated to body mass. It has been hypothesized that hyper- and hypo-sensitivity to ghrelin would in some way play a role in exaggerated or insufficient food intake, respectively. Along this line stands the more remarkable successful impact of gastric by-pass for treatment of obesity as a consequence of extremely low ghrelin secretion. It is however necessary to consider that the complexity of a neuro-humoral network dedicated to control food intake and energy balance does not depend on one major agent. Preliminary results obtained with ghrelin knock-out mice do not support a major role for ghrelin in the pathogenesis of weight disorders. Moreover, although ghrelin and its synthetic analogues are able to increase appetite and food intake, it is still unknown what happens in terms of adaptation and desensitization during chronic exposure to these molecules; it is at present unpredictable if ghrelin analogues will be able to represent a notable exception to what happens with other molecules in term of desensitization. Again, it is definitely clear and widely accepted that obesity and anorexia cannot simply be explained as a consequence of exaggerated or deficient appetite which is easily manipulated.

Other actions of ghrelin suggest that synthetic agonists and antagonists would have a useful impact on specific functions. In this context the impact of GHS on cardiovascular functions is receiving great attention. In light of the ability of ghrelin and synthetic GHS to directly improve cardiac performance, protect from ischemia and cardiomyocyte apoptosis, it is clear that another therapeutic dream would be generated.

However, the most important value in the ghrelin story now is the opportunity to better understand physiology as a function of a new factor; this in turn implies a more appropriate understanding of the role of the ghrelin system in some pathophysiological conditions.

Genetic studies as well as animal and human models of severe ghrelin hypo- and hyper-secretion will play a critical role in the understanding of ghrelin relevance in pathophysiological states.

The study of ghrelin secretion regulation is a critical point depending on the development of more appropriate methods of measurement and evaluation. In this context, at present, the following points await clarification: 1) the stomach is the major source of ghrelin, gastrectomy strongly decreases circulating ghrelin levels but where does the remaining 30% of ghrelin come from? Moreover, why does progressive recovery of circulating ghrelin levels after gastrectomy occur?; 2) we already know that there are major factors exerting inhibitory influences on ghrelin secretion, i.e., glucose, insulin, somatostatin, etc., but we do not know of stimulatory factors: we simply know that fasting and energy restriction stimulate ghrelin levels in agreement with the evidence that they are inhibited by meals and overfeeding; 3) we know that circulating ghrelin levels are mostly represented by the unacylated form that is devoid of any endocrine action; what about the regulation of the endocrinologically active acylated ghrelin?

What are the critical questions that need to be answered in order to better understand the ghrelin field? Among so many open questions, there are some we have not yet considered.

If the question is "How many known GHS-R and ligands are there?", the answer is GHS-R type 1a (GHS-R1a) and GHS-R type 1b (GHS-R1b). We have some good knowledge about distribution and functions of GHS-R1a and we assume it is the specific receptor for its natural ligand ghrelin provided that the molecule is acylated in serine 3; in fact, nonacylated ghrelin is unable to recognize this receptor. We also know that des-Gln14-acyl-ghrelin is another natural ligand of the GHS-R1a, that in turn, is also surprisingly bound and activated by adenosine. On the other hand, despite some homology between GHS-R and motilin receptor as well as between ghrelin and motilin, motilin does not bind the GHS-R1a. An important question is, on the other hand, "What is the role of the widespread distributed GHS-R1b?" This aspect is totally unknown at present.

Two other questions are probably of critical importance: 1) how many unknown GHS-R are there?; 2) is ghrelin "the" or "a" natural ligand of the GHS-R? i.e., are there other natural GHS-R ligands?

Let's start from the last question. Concerning GHS-R1a ligands, it has already been anticipated that des-Gln14-acyl-ghrelin and adenosine are other natural ligands of this ghrelin receptor. Interestingly, it has also been reported that cortistatin, a natural somatostatin analogue binding with high affinity all somatostatin receptors, also binds the GHS-R1a that in turn, is not recognized by somatostatin itself. The existence of a specific cortistatin receptor had already been hypothesized, taking into account that cortistatin also has some actions that are not shared by somatostatin. In agreement with this hypothesis, a previously orphan G-protein-coupled receptor named MrgX2 has been recently described that is able to selectively bind cortistatin but not somatostatin, suggesting that this could be one receptor selective for cortistatin only. Thus it is intriguing to understand what the meaning of cortistatin binding to the ghrelin receptor is.

Some observations support the possible existence of unknown GHS-R. There is a considerable difference between the distribution of GHS-R1a expression, the specific binding of labeled ghrelin or nonpeptidyl GHS and the specific binding of labeled peptidyl GHS in central and peripheral, endocrine and nonendocrine tissues. Again, it is assumed that unacylated ghrelin is devoid of biological activities in agreement with evidence that it does not bind the GHS-R1a. However, the existence of specific binding for labeled acyl-ghrelin has already been demonstrated; this binding is displaced by acyl- or nonacyl-ghrelin as well as by synthetic GHS in some tissues and cell lines. Notice that this binding is coupled with sharing the same activities of acyl ghrelin and synthetic GHS such as antiapoptotic action on cardiomyocytes and endothelial cells as well as antiproliferative effect on breast carcinoma cell lines. So, unacylated ghrelin possesses some biological actions that are probably mediated by a specific receptor that recognizes ghrelin independently of its acylation, i.e., a non GHS-R1a subtype recognizing synthetic GHS as well.

It has also been demonstrated that ghrelin as well as nonpeptidyl GHS has impact on glucose metabolism and insulin secretion, that are unaffected by peptidyl GHS. Indeed, ghrelin is expressed and synthesized within the endocrine pancreas in physiological and pathological conditions; however, GHS-R1a expression is not obviously present within pancreatic islets. This evidence would suggest that the peripheral metabolic actions of ghrelin are mediated by another GHS-R subtype that does not recognize peptidyl GHS.

Finally, GHS-R1a expression has been demonstrated in the cardiovascular system but at this level peptidyl GHS possess specific binding sites that are not recognized by either acylated or unacylated ghrelin as well as by nonpeptidyl GHS; therefore, this receptor is not a ghrelin receptor. Interestingly, both acylated and unacylated ghrelin and synthetic

GHS possess common cardiotropic actions, e.g., improving cardiac performance and exerting antiapoptotic actions. However, peptidyl GHS but not ghrelin exert a protective action against cardiac ischemia and this action is probably mediated by the above-mentioned specific receptor. This receptor has been shown to be identical to rat CD36, a multifunctional glycoprotein, that is expressed by cardiomyocytes and endothelial cells. It has also been shown that the activation of CD36 in perfused hearts by peptidyl GHS (but not ghrelin) is able to modulate coronary perfusion pressure in a dose-dependent manner. Moreover, this action is lacking in hearts from CD36-null mice and hearts from spontaneous hypertensive rats genetically deficient in CD36. This evidence implies that the natural ligand for this receptor is still unknown.

This is probably enough to definitely confuse the reader at the end of this book on ghrelin. Indeed, we are faced with a complex and intriguing matter that seems to project ghrelin from neuroendocrinology and metabolism to internal medicine. The most reasonable conclusion is that ghrelin has been shown to be much more than just a natural GHS and/or an orexigenic factor, and that this oversimplification can now be abandoned.

INDEX